Advances in Industrial Control

Springer
London
Berlin
Heidelberg
New York
Barcelona
Hong Kong
Milan
Paris
Santa Clara
Singapore
Tokyo

Other titles published in this Series:

Energy Efficient Train Control
P.G. Howlett and P.J. Pudney

Hierarchical Power Systems Control: Its Value in a Changing Industry
Marija D. Ilic and Shell Liu

System Identification and Robust Control
Steen Tøffner-Clausen

Genetic Algorithms for Control and Signal Processing
K.F. Man, K.S. Tang, S. Kwong and W.A. Halang

Advanced Control of Solar Plants
E.F. Camacho, M. Berenguel and F.R. Rubio

Control of Modern Integrated Power Systems
E. Mariani and S.S. Murthy

Advanced Load Dispatch for Power Systems: Principles, Practices and Economies
E. Mariani and S.S. Murthy

Supervision and Control for Industrial Processes
Björn Sohlberg

Modelling and Simulation of Human Behaviour in System Control
Pietro Carlo Cacciabue

Modelling and Identification in Robotics
Krzysztof Kozlowski

Spacecraft Navigation and Guidance
Maxwell Noton

Robust Estimation and Failure Detection
Rami Mangoubi

Adaptive Internal Model Control
Aniruddha Datta

Price-Based Commitment Decisions in the Electricity Market
Eric Allen and Marija Ilić

Compressor Surge and Rotating Stall
Jan Tommy Gravdahl and Olav Egeland

Radiotherapy Treatment Planning
Oliver Haas

Feedback Control Theory For Dynamic Traffic Assignment
Pushkin Kachroo and Kaan Özbay

Control Instrumentation for Wastewater Treatment Plants
Reza Katebi, Michael A. Johnson and Jacqueline Wilkie

Rick Lind and Marty Brenner

Robust Aeroservoelastic Stability Analysis

Flight Test Applications

With 70 Figures

Springer

Rick Lind, PhD, MS, BS, BS

Marty Brenner, MS, MS, BS
Dryden Flight Research Center
MS 4840 D/RS
CA 93523
USA

ISBN-13:978-1-4471-1215-0 Springer-Verlag London Berlin Heidelberg

British Library Cataloguing in Publication Data
Lind, Rick
Robust aeroservoelastic stability analysis : flight test applications. - (Advances in industrial control)
1.Flight control 2.Aeroelasticity - Stability
I.Title II.Brenner, Marty
629.1'326
ISBN-13:978-1-4471-1215-0

Library of Congress Cataloging-in-Publication Data
Lind, Rick, 1968-
Robust aeroservoelastic stability analysis : flight test applications / Rick Lind and Marty Brenner.
p. cm. -- (Advances in industrial control)
Includes bibliographical references (p.).
ISBN-13:978-1-4471-1215-0 e-ISBN-13:978-1-4471-0849-9
DOI: 10.1007/978-1-4471-0849-9

1. Stability of airplanes--Mathematical models.
2. Aeroelasticity. I. Brenner, Marty, 1955- . II. Title.
III. Series.
TL574.S7L47 1999
629.132'362'015118--dc21 98-52346
CIP

Softcover reprint of the hardcover 1st edition 1999

Typesetting: Camera ready by authors

69/3830-543210 Printed on acid-free paper

Advances in Industrial Control

Professor Dr -Ing M. Thoma
Institut für Regelungstechnik
Universität Hannover
Appelstr. 11
30167 Hannover
Germany

Professor H. Kimura
Department of Mathematical Engineering and Information Physics
Faculty of Engineering
The University of Tokyo
7-3-1 Hongo
Bunkyo Ku
Tokyo 113
Japan

Professor A.J. Laub
College of Engineering - Dean's Office
University of California
One Shields Avenue
Davis
California 95616-5294
United States of America

Professor J.B. Moore
Department of Systems Engineering
The Australian National University
Research School of Physical Sciences
GPO Box 4
Canberra
ACT 2601
Australia

Dr M.K. Masten
Texas Instruments
2309 Northcrest
Plano
TX 75075
United States of America

Professor Ton Backx
AspenTech Europe B.V.
De Waal 32
NL-5684 PH Best
The Netherlands

SERIES EDITORS' FOREWORD

The series Advances in Industrial Control aims to report and encourage technology transfer in control engineering. The rapid development of control technology impacts all areas of the control discipline. New theory, new controllers, actuators, sensors, new industrial processes, computer methods, new applications, new philosophies,...., new challenges. Much of this development work resides in industrial reports, feasibility study papers and the reports of advanced collaborative projects. The series offers an opportunity for researchers to present an extended exposition of such new work in all aspects of industrial control for wider and rapid dissemination.

The high performance control systems applications in aerospace and astronautics almost have a tradition of exploiting the most advanced control theoretical developments first. The optimal control and filtering paradigm associated with the names of Kalman, Bucy, Anderson and Moore found application in the astronautics of the 1960's and 1970's.

At the beginning of the 1980's, control theory moved on to robustness, singular values and mu-analysis. This new work was associated with the names of Zames, Doyle, Glover, Balas among others. The Advances in Industrial Control monograph series have published several volumes over the years which have archived the applications experience garnered from applying robust control to the aerospace sector problems. Rick Lind and Marty Brenner add to this set with their volume on robust aeroservoelastic stability. This volume reports the application of the structured singular value to aeroelastic and aeroservoelastic aerospace problems. A complete and systematic presentation is made of this area so that the monograph is essentially self-contained. A strong motivation for the work was the cost saving obtained from reducing the number of test flights needed to probe and determine safe flight envelopes.

The monograph is recommended for all specialists in the aerospace sector and also for the general reader seeking an insight into the practical application of the mu-analysis method.

M.J. Grimble and M.A. Johnson
Industrial Control Centre
Glasgow, Scotland, UK

ACKNOWLEDGEMENTS

The authors would like to thank the Structural Dynamics group at NASA Dryden Flight Research Center for their invaluable assistance in developing the material for this text. Leonard Voelker was instrumental in formulating the μ method by continuously providing lessons about the physics of flutter and how to interpret μ in terms of this phenomenon. Larry Freudinger shared his vast knowledge of data analysis and helped to develop the concepts of uncertainty for aeroservoelastic models. Dave Voracek gave in-depth information about the workings of an F/A-18 and demonstrated methods of flight flutter testing that could be improved by applying the μ method. Roger Truax and Tim Doyle generated the finite element model of the F/A-18 and computed nominal stability margins using classical methods. Mike Kehoe suggested many research avenues to address practical concerns of the aeroelasticity community and strengthen the applicability of the μ method to flight test programs. Also, Kajal Gupta deserves significant recognition for his efforts at developing the high fidelity tool for aeroservoelastic modeling and analysis that we relied upon to investigate stability of the F/A-18.

The authors would also like to acknowledge several other people that have contributed to the making of this text. Gary Balas has taught us many lessons about robust stability and the concept of μ in addition to providing insight into uncertainty modeling. Kari Appa was a dedicated and inspiring mentor in the area of aeroelasticity and flight dynamics. Tom Strganac and Andy Kurdila developed an aeroservoelastic testbed and generously provided models as an example with which to compute robust stability margins. Additionally, we would like to thank Rudy Yurkovich, Ken Griffin, Mordechay Karpel and Eli Livne for their helpful comments and support with regard to using μ for flight test programs.

CONTENTS

NOMENCLATURE

a_x, a_y, a_z	body axis perturbation accelerations
b	wing span
b_t	total actuator damping
$\bar{b}$	half mean aerodynamic chord
$\bar{c}$	mean aerodynamic chord
g	aeroelastic damping
$\bar{g}$	gravitational acceleration
k	reduced frequency
k_T	total control surface stiffness
k_0, b_0, m_0	aero stiffness, aero damping, and control surface mass
k_1, b_1, m_1	modal stiffness, damping, and mass
n	number of actuator hydraulic systems
n_s	number of states
n_i	number of inputs
n_o	number of outputs
n_Q	number of states in unsteady aerodynamic force matrix
n_w	number of feedback signals from uncertainty to plant
n_z	number of feedback signals from plant to uncertainty
p, q, r	perturbation roll, pitch, and yaw rate
$\bar{q}$	dynamic pressure
$\bar{q}_o$	nominal value of dynamic pressure
$\bar{q}_{nom}$	dynamic pressure associated with nominal stability analysis
$\bar{q}_{rob}$	dynamic pressure associated with robust stability analysis
r	position vector
s	Laplace variable
s_i	actuator controller state
t	time
u	input signal
y	output signal
w	feedback signal from uncertainty to plant
x	state vector
$\dot{x}$	derivative of x with respect to time
x_h	actuator ram moment arm
x_p	displacement of piston relative to main ram body
x_v	displacement of main control valve
x_0	displacement of control surface mass
x_1	displacement of structural modal mass
z	feedback signal from plant to uncertainty
$\bar{z}$	discrete variable
A_e	aerodynamic influence coefficient matrix
A_i	i^{th} term in Roger's form of unsteady aerodynamic forces

A_p	main actuator ram piston area
A_x, A_y, A_z	body axis accelerations
A_Q	state matrix of unsteady aerodynamic force matrix
B_e	effective stiffness correction matrix
B_Q	input matrix of unsteady aerodynamic force matrix
C_c	cross piston leak flow coefficient
C_l, C_m, C_n	coefficients of roll, pitch, and yaw moment
C_p	actuator fluid flow pressure constant
C_v	main valve flow coefficient
C_D, C_L, C_Y	coefficients of drag, lift, and side force
C_Q	output matrix of unsteady aerodynamic force matrix
D_Q	feedthrough matrix of unsteady aerodynamic force matrix
C	damping matrix
D	scaling matrix
$\mathcal{D}$	set of scaling matrices
F_e	externally applied actuator test force
$F_u(.,.)$	upper loop linear fractional transformation in Definition 3.1.1
$F_l(.,.)$	lower loop linear fractional transformation in Definition 3.1.2
G	scaling matrix
$\mathcal{G}$	set of scaling matrices
K	stiffness matrix
I_n	identity matrix of dimension n
M	mass matrix
P	transfer function operator
$\mathcal{P}$	set of system operators
P_j	pressure difference across actuator ram piston
P_{ij}	transfer function element of P
P_1, Q_1, R_1	reference (trim) roll, pitch, and yaw rate
Q	unsteady aerodynamic force matrix
$\bar{Q}$	Q without the rigid air loads
S	state-space system operator
$\bar{S}$	reference area of wing
T	sample time for digital controller
U_1, V_1, W_1	reference (trim) velocities in body-fixed frame of reference
W	matrix scaling uncertainty operators
$W_{\bar{q}}$	matrix scaling feedback perturbation to dynamic pressure

DEI	Dynamic Engineering Incorporated
KEAS	knots of equivalent airspeed
LFT	linear fractional transformation
LMI	linear matrix inequality
LTI	linear time-invariant
LTV	linear time-varying
SRA	Systems Research Aircraft

α	angle of attack
α_1	trim angle of attack
$\alpha(\omega)$	equivalent downwash
β	angle of sideslip
β_i	aerodynamic lag terms
δ	uncertainty scalar
δ_d	actuator command
$\delta_{\overline{q}}$	uncertainty perturbation on dynamic pressure
Δ	uncertainty matrix
$\boldsymbol{\Delta}$	set of uncertainty matrix operators
η	generalized coordinate
θ, ϕ, ψ	perturbation pitch, roll, and yaw angles
Θ_1, Φ_1, Ψ_1	reference (trim) pitch, roll, and yaw angles
Γ	stability margin expressed as difference in flight condition
Γ_i	inertial to body-fixed coordinate transformation matrices
$\jmath$	imaginary unit of $\sqrt{-1}$
$\Lambda, \tilde{\Lambda}, \Lambda_i, \tilde{\Lambda}_i$	inertial to body-fixed coordinate transformation matrices
μ	structured singular value
Π	stability margin expressed as percentage of flight condition
ρ	spectral radius
σ	singular value
$\overline{\sigma}$	maximum singular value
τ	system time delay
τ_i	actuator filter time constants
$\Phi_r, \Phi_e, \Phi_\delta$	rigid, elastic, and control mode shape matrices
ω	frequency in units of radians per second
Ω	set of frequency points
$\mathbf{C}^{n\times m}$	space of complex valued matrices of dimension n by m
$\mathbf{R}^{n\times m}$	space of real valued matrices of dimension n by m
$\mathcal{H}_2$	Hardy space in Definition 2.1.8
$\mathcal{H}_\infty$	stable rational transfer functions in Definition 2.2.3
$\mathcal{RH}_\infty$	stable rational transfer functions in Definition 2.2.4
$\mathcal{L}_\infty$	Lesbegue space of transfer functions in Definition 2.2.2
$\mathcal{L}_2$	frequency domain Lebesgue space in Definition 2.1.7
$\mathcal{L}_2(-\infty, \infty)$	time-domain Lebesgue space in Definition 2.1.3
$\mathcal{L}_2[0, \infty)$	time-domain Lebesgue space in Definition 2.1.4
$\mathcal{S}$	set of signals in Definition 2.1.1
$\mathcal{S}_{\jmath\omega}$	set of frequency domain signals in Definition 2.1.5
$\lVert x(t)\rVert_2$	2-norm of time domain signal in Definition 2.1.2
$\lVert f(\jmath\omega)\rVert_2$	2-norm of frequency domain signal in Definition 2.1.6
$\lVert P\rVert_\infty$	$\mathcal{H}_\infty$-norm induced by $\mathcal{L}_2$ signals in Definition 2.2.1

CHAPTER 1
INTRODUCTION

Aeroservoelasticity considers the interaction between aerodynamics, inertial, structural, actuation, and control system dynamics. The stability properties of the aeroservoelastic dynamics must be investigated to determine a flight envelope that is clear of instabilities for new aircraft designs or new configurations of current aircraft. Traditionally filters have been included with control systems to eliminate any aeroservoelastic interactions; however, the stringent performance objectives and enhanced structural flexibility of modern aircraft are making the analysis of aeroservoelastic dynamics increasingly important.

Critical aeroservoelastic instabilities are denoted as the operating points closest to the flight envelope at which an instability is encountered. The concept of closeness is formally defined in terms of a critical pressure that describes the value of dynamic pressure at a critical instability when considering points along a constant Mach trajectory. Obviously different parameters such as a critical speed can be defined since a unique equivalent airspeed is associated with each dynamic pressure at a particular Mach number; however, Definition 1.0.1 will be used as the basic concept associated with a critical instability.

Definition 1.0.1 *A critical aeroservoelastic pressure is the smallest value of dynamic pressure for which an aircraft at a particular Mach number experiences an aeroservoelastic instability.*

The concept of an aeroservoelastic stability margin is defined to indicate the distance between this critical pressure and a reference point. A common stability margin, Γ, measures distance by the difference between values of dynamic pressure for the critical and reference points. Another margin, Π, considers a percentage difference between equivalent airspeeds at the critical pressure and the reference point. These margins are encompassed by the general concept given in Definition 1.0.2.

Definition 1.0.2 *An aeroservoelastic stability margin relates a measure of distance between the flight conditions associated with the critical pressure and a reference point.*

The result of an aeroservoelastic stability analysis is the determination of a flight envelope within which the aircraft may safely operate. Traditionally this determination has been based on classical stability concepts associated with a theoretical model. In particular, guidelines have been established that require a minimum level of aeroservoelastic stability margin as measured from the edge of the acceptable flight envelope.

Aeroservoelasticity has not traditionally been a dominant research area for aircraft; rather, the field of aeroelasticity has received considerable attention. Aeroelasticity considers the interaction of aerodynamic, inertial, and structural forces and so is actually a subset of aeroservoelasticity [13]. In particular, the instability of flutter has been studied with great detail.

The mechanisms associated with flutter may be different from those associated with aeroservoelastic instabilities and so should be considered separately. A critical pressure for aeroelastic dynamics is the smallest flight condition at which an aircraft experiences flutter. The concept of an aeroelastic stability margin, or flutter margin, is given in Definition 1.0.3 as the aeroelastic equivalent to Definition 1.0.2 based on a critical flutter pressure.

Definition 1.0.3 *A flutter margin relates a measure of distance between the flight condition associated with the flutter pressure and a reference point.*

The aeroelasticity community has identified developing improved methods for characterizing aeroservoelastic and flutter margins as a vital research area [30]. Theoretical predictions that are based on a computational model may be misleading because of approximations and modeling errors. Similarly, experimental predictions that are based on flight data analysis may also be misleading because of incorrect assumptions and erroneous estimations caused by low signal to noise ratios.

Flight testing for envelope expansion incurs dramatic time and cost because of the potentially misleading information provided by traditional methods [66]. In particular, flutter instabilities often exhibit an explosive behavior that causes a sudden change in stability despite only a small change in flight condition. Thus, small errors in predicted margins could have grave consequences for aircraft and crews operating near aeroservoelastic instabilities.

Computational methods to predict instabilities are continually being improved as result of better aerodynamic and structural models; however, these models will always have errors. Approaches have been developed to try and estimate some level of confidence in computational margins by evaluating robustness of a stability margin with respect to the changes in the model. A high level of robustness is considered a measure of confidence since modeling errors may not greatly affect the critical pressures. An algorithm is developed to compute the most critical flutter margin with respect to first order perturbations in a model [7]. This method considers only parametric perturbations and may be computationally expensive. A robust control framework is adopted using a feedback structure to relate the structural model and the aerodynamic model [8]. This approach uses highly conservative robustness conditions with an uncertainty structure that may not be physically meaningful. A similar approach is adopted allowing unmodeled dynamics and higher order parametric perturbations based on series expansion [9]. Statistical approaches are also considered to formulate a flutter probability measure [103, 109]. These approaches will converge to a robustness indicator using Monte Carlo simulations but the computation time may be prohibitive for complex systems.

The robustness measures for these perturbation and statistical approaches are suspect since no global guarantees can be made as to perturbations not explicitly considered by the minimization algorithms or the Monte Carlo simulations.

Flight data presents several challenges for methods that attempt to process this data and estimate stability margins; however, this data provides the only information about the true aircraft and so should be considered in any stability analysis. Parameter estimation algorithms are developed to directly identify an aeroelastic model from the flight data [46, 95]. The accuracy of the resulting model may deteriorate as the complexity and number of degrees of freedom of the system increases along with signal to noise ratios decreasing from optimal wind tunnel conditions to realistic flight levels. Modal filtering is introduced in association with parameter estimation algorithms to simplify analysis by decoupling the system into a set of first order responses [118, 119]. This type of filtering does not guarantee robustness and may not perform well for systems with many closely spaced modal natural frequencies which cross and shift as flight conditions change.

There are several advantages and disadvantages to these suggested methods for computational and experimental derivation of stability margins. Computational methods are advantageous because they use well-developed models and algorithms; however, these models are only approximations and even estimates of sensitivities may be based on incorrect assumptions of modeling errors. Conversely, experimental methods use flight data from the physical aircraft and thus reflect properties of the true dynamics; however, these methods require high signal to noise ratios that are often not measured in flight.

The μ method has been developed to use both a computational model and flight data and thereby combine the advantages of each method while avoiding the drawbacks [77]. This method uses a robust stability framework to relate a theoretical aeroservoelastic model with a set of norm-bounded operators that describe modeling errors and uncertainties. A multivariable robust stability measure known as the structured singular value, μ, computes a stability margin that is robust to the included level of modeling errors. A robust aeroservoelastic stability margin is thus computed as the largest increase in dynamic pressure for which the model remains stable despite variations admitted by the set of potential modeling errors.

Flight data is easily incorporated into the μ analysis procedure. The modeling errors are determined by comparing transfer functions obtained by flight data with transfer functions predicted by the analytical model. The norm bound on the uncertainty operators is chosen based on these observed errors. A model validation condition is used to ensure the resulting operator set is sufficient to account for multiple data sets without being excessively conservative. In this sense, a *worst-case* stability margin is computed that directly accounts for flight data.

This method is inherently different from traditional algorithms based on p-k methods or parameter identification and robustness approaches. The μ method utilizes information from both an analytical model and flight data while traditional approaches use only one of these as demonstrated in Figure 1.1. Methods which utilize only an analytical model may be inaccurate while methods which utilize only the flight data may fail if that data is of poor quality. The μ method uses the flight data to improve the analytical model by the addition of uncertainty operators so poor quality flight data will merely increase the difficulty of obtaining a reasonable uncertainty description resulting in a small set $\mathbf{\Delta}$. The robust margins will be similar to the nominal margins in this case which makes intuitive sense since any information obtained from the data should only enhance the plant model and improve the accuracy of the flutter margin.

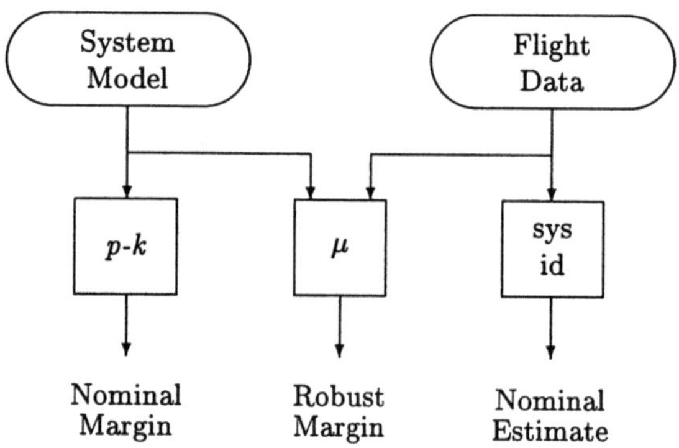

Fig. 1.1: Information Flowchart for the μ Method and Traditional Computational and Experimental Methods to Compute Aeroservoelastic Stability Margins

The concept of computing robustness in flutter margins is recognized for its importance and is considered a state of the art research area in aeroelasticity [60]. Informal measures of robustness are not necessarily useful since they provide no guarantees as to the system stability. The μ method is based on operator theory and provides a well defined concept of robustness with a clear set of guarantees as to the stability properties of the system.

Also, the μ method allows consideration of a rich set of modeling errors with direct physical interpretations. Traditional techniques compute robustness with respect to a theoretical estimate of modeling errors that may be too conservative or overly optimistic. The μ method is able to use flight data to determine these errors by employing model validation algorithms which are developed specifically for comparing data with uncertain models in the μ framework. Thus, the resulting stability margins directly account for the flight data when determining robust flutter pressures.

The μ method may greatly reduce the risk and cost associated with flight flutter testing for envelope expansion since the aeroservoelastic stability margins may be more accurately and confidently identified. Flight data from a safe flight condition may be evaluated to determine the amount of modeling error and robust flutter margins can be computed to indicate future safe test points. This method would require less flight time at a reduced set of flight test points as compared to the procedure required to slowly expand the envelope when the flutter margins are not confident. On-line implementations of the μ method have been used to develop the concept of a flutterometer that would indicate a distance to flutter as the envelope is expanded and thus increase efficiency of a flight flutter test [72].

This text details the μ method for analyzing aeroservoelastic and flutter instabilities. In particular, the models that are used to represent the aircraft in the μ framework are explicitly derived from the general aeroelastic equations of motion. Several types of uncertainty operators are presented to describe typical modeling errors and shown how they can be associated with the nominal models. Algorithms are given to demonstrate how robust aeroservoelastic and flutter stability margins are computed and how they can be interpreted in terms of aircraft dynamics and flight safety.

Several applications of the μ method to flight systems are presented to demonstrate the concept of robust aeroservoelastic stability analysis. These applications show how flight data can be used to indicate uncertainties that describe potential modeling errors. Furthermore, they clearly show that the uncertainties can have significant effects on the resulting stability margins even for systems that are not considered overly susceptible to aeroservoelastic instabilities.

CHAPTER 2
ROBUST STABILITY

2.1 Signals

A common definition of a signal is a Lebesgue measurable function that maps the space of real numbers $\mathbf{R}$ into $\mathbf{R}^n$. A space of such signals is denoted $\mathcal{S}$.

Definition 2.1.1 *The space of signals which are Lebesgue measurable functions is $\mathcal{S}$.*

$$\mathcal{S} = \{f : \mathbf{R} \to \mathbf{R}^n\}$$

Analog measurements $x(t)$ of physical systems are real-valued vector functions of the real parameter t representing time and thus are valid members of the space of signals such that $x(t) \in \mathcal{S}$. Values of the time parameter, which are often arbitrarily numbered as a distance from some reference point, actually extend to positive and negative infinity; thus, stability for physical systems must ensure stability for all values of time. A time-domain 2-norm is defined as a measure of size, or energy, for time-domain signals that considers all time.

Definition 2.1.2 *The energy of $x(t) \in \mathcal{S}$ is measured by $\|x\|_2$.*

$$\|x(t)\|_2 = \left(\int_{-\infty}^{\infty} |x(t)|^2 dt \right)^{\frac{1}{2}}$$

One characteristic of a stable system is that only finite-energy output signals are generated in response to finite-energy input signals. Signals with finite energy are known as square integrable since the integral of the square of the signal is finite. The Lebesgue space of square-integrable signals is defined as $\mathcal{L}_2(-\infty, \infty)$. This space is also referred to as the infinite-horizon Lebesgue 2-space to denote the norm utilizes an integral over infinite time.

Definition 2.1.3 *The space $\mathcal{L}_2(-\infty, \infty)$ consists of square-integrable time-domain signals.*

$$\mathcal{L}_2(-\infty, \infty) = \{x(t) \ : \ x \in \mathcal{S}, \|x(t)\|_2 < \infty\}$$

Signals associated with physical systems are only known for values of time greater than the time at which measurements are started. Stability analysis and norm computations using these signals can not utilize properties of the signal prior to the starting time since no information is known so the traditional method of characterizing these signals is to assume the values are identically zero for all times before the starting time. The time value at which measurements are started may be chosen without loss of generality and is usually chosen to be $t = 0$. The space $\mathcal{L}_2[0, \infty)$ is defined as a subset of $\mathcal{L}_2(-\infty, \infty)$ to emphasize such signals.

Definition 2.1.4 *The space $\mathcal{L}_2[0, \infty) \subset \mathcal{L}_2(-\infty, \infty)$ consists of square-integrable time-domain signals which are identically zero for all $t < 0$.*

$$\mathcal{L}_2[0, \infty) = \left\{ x(t) \; : \; \int_0^{\infty} \|x(t)\|^2 dt < \infty, x(t) = 0 \text{ for all } t < 0 \right\}$$

A similar space $\mathcal{L}_2(-\infty, 0] \subset \mathcal{L}_2(-\infty, \infty)$ is defined for signals $x(t)$ that are assumed to begin at $t = -\infty$ and are identically zero for all times $t > 0$. The integral to compute the energy for elements of this space can thus be restricted to consider $t = -\infty$ until $t = 0$.

Frequency-domain signals are often considered in stability analysis but do not fall into the set $\mathcal{S}$ of signals. These signals $f(\jmath\omega)$ are complex-valued functions of the imaginary unit $\jmath = \sqrt{-1}$ and the real variable ω that represents frequency in radians per second. The set $\mathcal{S}_{\jmath\omega}$ is defined for frequency-domain signals with a complex conjugate condition to reflect properties of $\jmath$.

Definition 2.1.5 *The space of frequency-domain signals is $\mathcal{S}_{\jmath\omega}$.*

$$\mathcal{S}_{\jmath\omega} = \{ f(\jmath\omega) : \jmath\mathbf{R} \to \mathbf{C}^n \text{ with } f^*(\jmath\omega) = f'(-\jmath\omega) \}$$

A 2-norm value is also formulated to compute a measure of energy.

Definition 2.1.6 *The 2-norm measures the energy of the signal $f(\jmath\omega) \in \mathcal{S}_{\jmath\omega}$.*

$$\|f\|_2 = \left(\frac{1}{2\pi} \int_{-\infty}^{\infty} f^*(\jmath\omega) f(\jmath\omega) d\omega \right)^{\frac{1}{2}}$$

A frequency-domain Lebesgue space is defined for finite-energy signals.

Definition 2.1.7 *The space $\mathcal{L}_2$ consists of frequency-domain signals with finite energy.*

$$\mathcal{L}_2 = \{ f(\jmath\omega) \; : \; f \in \mathcal{S}_{\jmath\omega}, \|f\|_2 < \infty \}$$

The spaces $\mathcal{L}_2(-\infty, \infty)$ and $\mathcal{L}_2$ are isomorphic Hilbert spaces under the appropriate inner products through the Fourier transform which means the spaces have equivalent algebraic properties. This relationship is used to simplify notation by rarely distinguishing between time-domain and frequency-domain signals except where the context does not make it clear. The notations for the 2-norm of time-domain and frequency-domain is also not distinguished since they are equivalent as demonstrated by Parseval's identity [86].

An important subspace of $\mathcal{L}_2$ is the Hardy space denoted as $\mathcal{H}_2$. This space contains the complex-variable functions which are analytic in the open right half of the complex plane and have finite 2-norm.

Definition 2.1.8 *The Hardy space $\mathcal{H}_2 \subset \mathcal{L}_2$ consists of the following functions.*

$$\mathcal{H}_2 = \{f(s) \,:\, f(s) \in \mathcal{L}_2 \text{ and } f(s) \text{ is analytic in } Re(s) > 0\}$$

The space $\mathcal{L}_2[0,\infty)$ is isomorphic to $\mathcal{H}_2$ through the Laplace transform; thus, members of this set can be interpreted as frequency-domain representations of finite-energy time-domain signals that are assumed to be identically zero for all times $t < 0$.

2.2 Systems

The systems to be considered are defined as operators mapping input signals to output signals. This implies that for every input signal, $u \in \mathcal{S}$, then there exists a corresponding output signal, $y \in \mathcal{S}$, that is related through $y = Pu$.

$$P : \mathcal{S} \to \mathcal{S}$$

In particular, systems are restricted to be linear, time-invariant operators that may be realized as state-space models.

$$\begin{bmatrix} \dot{x}(t) \\ y(t) \end{bmatrix} = \begin{bmatrix} A & B \\ C & D \end{bmatrix} \begin{bmatrix} x(t) \\ u(t) \end{bmatrix} \tag{2.1}$$

The signal $x(t) \in \mathbf{R}^{n_s}$ is the state vector, $u(t) \in \mathbf{R}^{n_i}$ is the input vector and $y(t) \in \mathbf{R}^{n_o}$ is the output vector. The system elements are defined such that $A \in \mathbf{R}^{n_s \times n_s}$ is the state update matrix, $B \in \mathbf{R}^{n_s \times n_i}$ determines how the input affects the states, $C \in \mathbf{R}^{n_o \times n_s}$ computes the outputs as a linear combination of states and $D \in \mathbf{R}^{n_o \times n_i}$ is the direct feedthrough from inputs to outputs. The operator $S = \{A, B, C, D\}$ denotes the time domain state-space system.

Linear, time-invariant state-space systems are commonly represented by transfer function operators. These functions $P(s)$ are complex valued matrices of the complex Laplace variable s. Such a transfer function matrix exists if and only if the state-space system is linear and time-invariant.

$$P(s) = \left[\begin{array}{c|c} A & B \\ \hline C & D \end{array}\right] = D + C(sI - A)^{-1}B \tag{2.2}$$

Stability must be considered over the infinite-horizon time lengths so the operators to be utilized must map $\mathcal{L}_2(-\infty,\infty)$ into $\mathcal{L}_2(-\infty,\infty)$. Properties of the Fourier transform relating $\mathcal{L}_2(-\infty,\infty)$ and $\mathcal{L}_2$ imply a state space system $S : \mathcal{L}_2(-\infty,\infty) \to \mathcal{L}_2(-\infty,\infty)$ is linear and time-invariant if and only if the associated transfer function matrix P is such that $y = Pu \in \mathcal{L}_2$ for any $u \in \mathcal{L}_2$. This leads to consideration of the gain for these signals.

$$\frac{\|y\|_2}{\|u\|_2} = \frac{\|Pu\|_2}{\|u\|_2} \tag{2.3}$$

This ratio of 2-norms will be finite if the system is stable. Properties of the 2-norm are used to derive a condition on the system operator P that is equivalent to analyzing if this ratio is finite. This condition is referred to as an induced norm since it results from consideration of signal norms associated with input and outputs of the operator. The $\mathcal{L}_2$-induced norm is defined as the $\mathcal{H}_\infty$-norm.

Definition 2.2.1 *Define the $\mathcal{H}_\infty$ norm for transfer function operators.*

$$\|P\|_\infty = \sup_\omega \overline{\sigma}(P(\jmath\omega))$$

A space of operators with finite $\mathcal{H}_\infty$-norm is denoted $\mathcal{L}_\infty$.

Definition 2.2.2 *The space $\mathcal{L}_\infty$ consists of systems with finite $\mathcal{H}_\infty$-norm.*

$$\mathcal{L}_\infty = \{P : \|P\|_\infty < \infty\}$$

Transfer functions of linear time-invariant systems are stable if and only if $z = Pw$ and $w \in \mathcal{H}_2$ implies $z \in \mathcal{H}_2$. This results from the Laplace isomorphism between $\mathcal{L}_2[0, \infty)$ and $\mathcal{H}_2$ space. These transfer functions are shown to be analytic in the open right-half complex plane with finite $\mathcal{H}_\infty$ norm. Define the space $\mathcal{H}_\infty$ to contain these operators.

Definition 2.2.3 *The space $\mathcal{H}_\infty$ consists of transfer functions of stable linear, time-invariant systems with finite $\mathcal{H}_\infty$-norm.*

$$\mathcal{H}_\infty = \{P : P \textit{ is analytic in } Re(s) > 0 \textit{ and } \|P\|_\infty < \infty\}$$

A subspace $\mathcal{RH}_\infty$ is often defined for rational elements.

Definition 2.2.4 *The space $\mathcal{RH}_\infty \subset \mathcal{H}_\infty$ consists of rational elements of $\mathcal{H}_\infty$.*

$$\mathcal{RH}_\infty = \{P : P \in \mathcal{H}_\infty \textit{ and } P \textit{ is rational}\}$$

Transfer function operators of linear, time-invariant state-space systems are rational functions of the Laplace transform variable s. These transfer functions $P \in \mathcal{RH}_\infty$ if and only if P is stable such that no poles lie in the closed right-half plane. The space $\mathcal{RH}_\infty$ which may appear to be a mathematical abstraction is thus shown to have a physical interpretation. $\mathcal{RH}_\infty$ is merely the operator theory representation of stable rational transfer functions.

2.3 Small Gain Theorem

Stability of a linear, time-invariant system is determined by the system poles or eigenvalue of the state matrix of a state-space realization. Robust stability in the $\mathcal{H}_\infty$ and μ frameworks is determined by considering an interconnection of stable operators. The basis for determining stability of these interconnections of operators is the *Small Gain Theorem* [142].

The Small Gain Theorem states that a closed-loop feedback system of stable operators is internally stable if the loop gain of those operators is stable and bounded by unity. Several formulations of the Small Gain Theorem are derived for various signals and systems. Theorem 2.3.1 presents the formulation used for developing robust aeroservoelastic stability margins [86].

Theorem 2.3.1 *[Small Gain Theorem] Given the feedback interconnection structure of Figure 2.1 for stable transfer function operators* $P, \Delta : \mathcal{L}_2 \to \mathcal{L}_2$ *with* $P, \Delta \in \mathcal{RH}_\infty$*, if the* $\mathcal{H}_\infty$*-norm of the loop gain is bounded by unity such that* $\|P\Delta\|_\infty < 1$*, then*

1. *the closed-loop system is well-posed and internally stable*
2. *a unique* $y, w \in \mathcal{L}_2$ *is associated with each* $u \in \mathcal{L}_2$

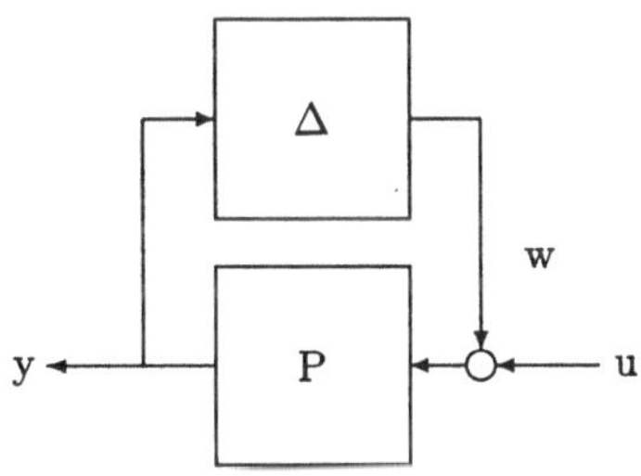

Fig. 2.1: Block Diagram for the Small Gain Theorem

The small gain theorem as presented in Theorem 2.3.1 is overly restrictive in the sense of requiring $P, \Delta \in \mathcal{RH}_\infty$. A more general small gain theorem is formulated for operators not restricted to lie in the subspace $\mathcal{RH}_\infty$ such that Theorem 2.3.1 is a special case of this general theorem [52]. The extended operator space in the general small gain theorem allows consideration of robustness for systems comprised of nonlinear and time-varying operators. The requirement of considering stable rational transfer function operators is explicitly stated in the theorem to emphasize the nominal aeroservoelastic system to be analyzed is assumed to be nominally stable and the robust stability margin is associated with a destabilizing perturbation to that nominal system.

The second condition in Theorem 2.3.1 is associated with the first condition guaranteeing a well-posed and stable system. This uniqueness condition may be understood by consideration of the solution y for the loop equations of Figure 2.1.

$$\begin{aligned} y &= P(u+w) \\ &= (I - P\Delta)^{-1}Pu \end{aligned} \tag{2.4}$$

The inverse term, $(I - P\Delta)^{-1}$, has a magnitude of infinity if the norm of $P\Delta$ is allowed to be unity. Such a condition would allow the norm of signal y to be infinite despite a bounded-norm input signal u. Restricting $||P\Delta||_\infty < 1$ ensures the inverse term exists and a unique finite-norm y is generated in response to a finite-norm u. The issue of well-posedness requires this condition to hold at $s = \infty$ and is automatically considered by the $\mathcal{H}_\infty$-norm.

2.4 Robust Stability

The small gain theorem may be directly utilized to analyze robust stability of a plant model with respect to a set of modeling errors by considering robustness to a set of perturbations [32]. These perturbations affect the nominal plant such that the true plant is assumed to correspond to some value of the perturbed nominal plant. Thus, the perturbations are used to describe uncertainty in the analytical plant model due to errors and unmodeled dynamics. Usually the exact value of the modeling error is not known but a bound, $\alpha > 0$, can be placed on the size of that error. Define the set $\boldsymbol{\Delta}$ of norm bounded operators describing these perturbations which affect the plant P through a feedback relationship.

$$\boldsymbol{\Delta} = \{\Delta : ||\Delta||_\infty \leq \alpha\} \tag{2.5}$$

The small gain theorem allows consideration of the entire set of possible modeling uncertainties as described by all $\Delta \in \boldsymbol{\Delta}$. The $\mathcal{H}_\infty$-norm of the loop gain can not be explicitly computed for these systems since there is an infinite number of loop gains $P\Delta$ generated by the set $\boldsymbol{\Delta}$; instead, the triangle inequality of norms can be utilized to generate a sufficient condition for robust stability of P.

$$||P\Delta||_\infty \leq ||P||_\infty ||\Delta||_\infty \tag{2.6}$$

A condition for robust stability of the closed-loop system is derived using the small gain theorem and (2.6). This condition is stated in Lemma 2.4.1.

Lemma 2.4.1 *The plant P is robustly stable to the set of uncertainty perturbations* $\mathbf{\Delta}$ *which enter the system as in Figure 2.1 with* $\|\Delta\|_\infty \leq \alpha$ *for all* $\Delta \in \mathbf{\Delta}$ *if*

$$\|P\|_\infty < \frac{1}{\alpha}$$

Note Lemma 2.4.1 presents a sufficient, but not necessary, condition for robust stability. The structured singular value, μ, is introduced in the next chapter as a less conservative measure of robust stability which is sufficient and necessary.

An excellent illustrative example is presented in Reference [52] to demonstrate the issue of robust stability. This example computes a robust stability condition for a simple system using classical arguments which is seen to be identical to the robust stability condition generated using the small gain theorem and Lemma 2.4.1. A similar example is presented here for the feedback interconnection in Figure 2.2.

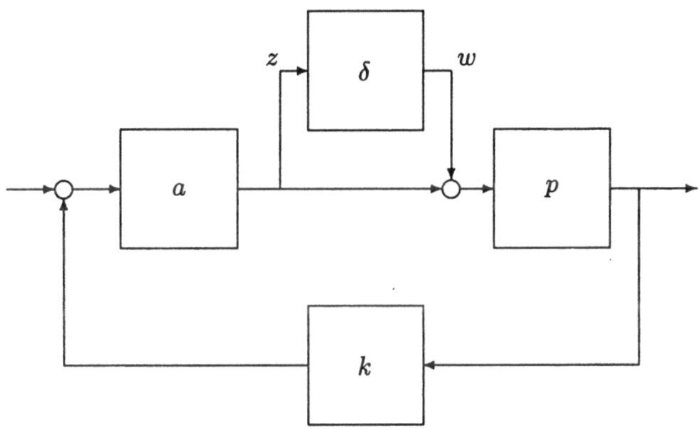

Fig. 2.2: Block Diagram with Uncertainty for the Example System

The single input and single output elements in the nominal system model are p which represents the plant dynamics, a which represents actuator dynamics, and k which represents a feedback controller. Each of the nominal system elements are stable transfer functions contained in $\mathcal{RH}_\infty$. There is a modeling error on the output of the actuator which is represented by a multiplicative uncertainty operator $\delta \in \mathcal{RH}_\infty$ on the output of a.

The transfer function from w to z may be computed.

$$z = \left(-(1+akp)^{-1}akp\right) w \tag{2.7}$$

Internal stability of the closed-loop feedback system is equivalent to stability of the feedback system shown in Figure 2.3 with the operator $g = -(1+akp)^{-1}akp$.

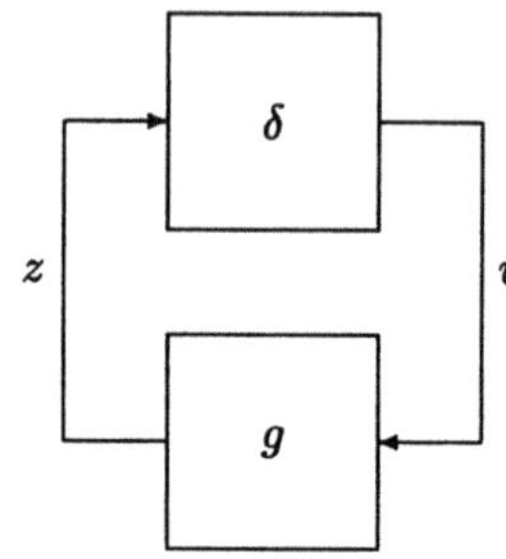

Fig. 2.3: Block Diagram for Robust Stability Analysis of the Example System using the Small Gain Theorem

Since the operators $\delta, g \in \mathcal{RH}_\infty$ are stable, the Nyquist criterion determines the closed-loop system is stable if and only if the Nyquist plot of δg does not encircle the -1 point. This stability condition is equivalent to the following norm condition.

$$\sup_\omega |g(\jmath\omega)\delta(\jmath\omega)| < 1 \tag{2.8}$$

This condition is an $\mathcal{H}_\infty$-norm condition on the loop gain $g\delta$. Thus, classical Nyquist arguments derive an $\mathcal{H}_\infty$-norm condition that is equivalent to the stability condition immediately formulated by applying the small gain theorem.

$$\text{closed-loop stability} \quad \Leftrightarrow \quad \|g\delta\|_\infty < 1$$

The error in the actuator command is unknown and possibly time-varying so the operator δ is used to allow consideration of a range of errors. Assume the actuator is weighted such that the range of errors is described by the set of perturbations $\|\delta\|_\infty < 1$. Lemma 2.4.1 is used to generate a condition that ensures the system is robustly stable to all actuators errors described by δ.

$$\text{closed-loop stability} \quad \Leftarrow \quad \|g\|_\infty < 1$$

CHAPTER 3

STRUCTURED SINGULAR VALUE: μ

3.1 Linear Fractional Transformations

The linear fractional transformation (LFT) is a common framework suitable for robust stability analysis using arguments based on the small gain theorem. An LFT is an interconnection of operators arranged in a feedback configuration. These operators may be constant matrices, time-domain state-space systems, or frequency-varying transfer functions. Consider a linear operator $P \in \mathbf{C}^{(o_1+o_2)\times(i_1+i_2)}$ which is partitioned into four elements.

$$P = \begin{bmatrix} P_{11} & P_{12} \\ P_{21} & P_{22} \end{bmatrix} \tag{3.1}$$

The linear fractional transformation, $F_u(P, \Delta)$, is defined as the interconnection matrix with the upper loop of P closed with the operator $\Delta \in \mathbf{C}^{i_1 \times o_1}$.

Definition 3.1.1 *Given $P \in \mathbf{C}^{(o_1+o_2)\times(i_1+i_2)}$ and $\Delta \in \mathbf{C}^{i_1\times o_1}$, define $F_u(P, \Delta)$ as the upper loop linear fractional transformation of P closed with Δ such that $y = F_u(P, \Delta)u$ as in Figure 3.1.*

$$F_u(P, \Delta) = P_{22} + P_{21}\Delta(I - P_{11}\Delta)^{-1}P_{12}$$

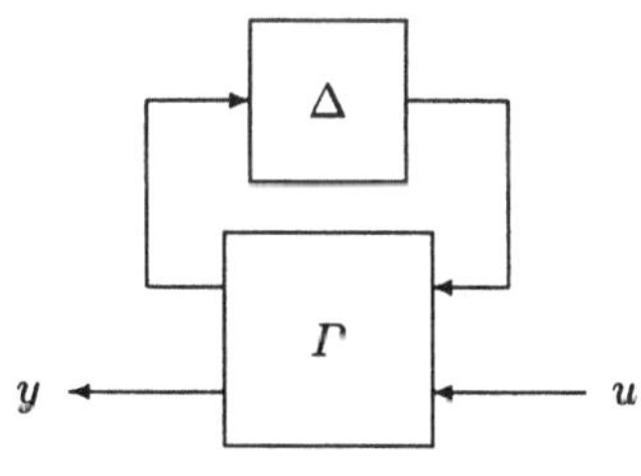

Fig. 3.1: Linear Fractional Transformation $F_u(P, \Delta)$

A similar LFT is defined as $F_l(P, \Delta)$ to represent the interconnection matrix of the lower loop of P closed with an operator $\Delta \in \mathbf{C}^{i_2\times o_2}$.

Definition 3.1.2 *Given $P \in \mathbf{C}^{(o_1+o_2)\times(i_1+i_2)}$ and $\Delta \in \mathbf{C}^{i_2\times o_2}$, define $F_l(P, \Delta)$ as the lower loop linear fractional transformation of P closed with Δ such that $y = F_l(P, \Delta)u$ as in Figure 3.2.*

$$F_l(P, \Delta) = P_{11} + P_{12}\Delta(I - P_{22}\Delta)^{-1}P_{21}$$

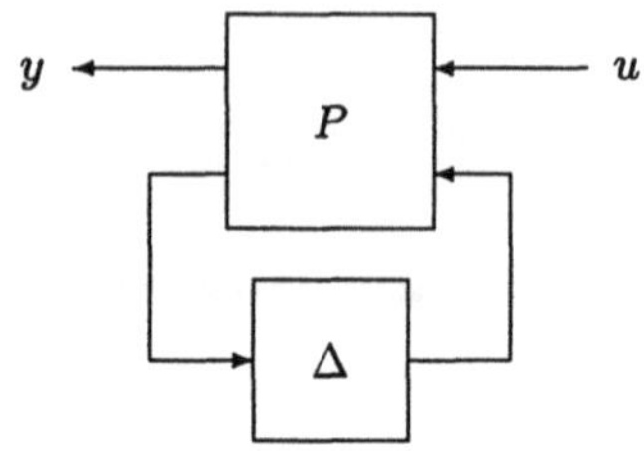

Fig. 3.2: Linear Fractional Transformation $F_l(P, \Delta)$

An example of an interconnection that is common in stability analysis is the representation of a time-dependent state-space system as a frequency-varying transfer function. Consider a state-space realization to compute a signal y in response to an input signal u with states x.

$$\begin{bmatrix} \dot{x} \\ y \end{bmatrix} = \begin{bmatrix} A & B \\ C & D \end{bmatrix} \begin{bmatrix} x \\ u \end{bmatrix} \tag{3.2}$$

An equivalent formulation to (3.2) is to compute y through the response of an LFT involving the Laplace variable s and a constant matrix whose entries are the state-space elements.

$$\begin{aligned} y &= \left(D + C(sI - A)^{-1}B\right) u \\ &= F_u\left(\begin{bmatrix} A & B \\ C & D \end{bmatrix}, \frac{1}{s}\right) \end{aligned} \tag{3.3}$$

The LFT is a useful framework for analyzing complex systems with many feedback and series interconnections of operators because of the interconnection principle given in Property 3.1.3. This property allows complex systems of several interconnected LFT's to be expressed as an equivalent single LFT. The operators of the new LFT are block structured with blocks composed of the individual operators of the LFT's from the original system.

Property 3.1.3 *Feedback and series interconnections of LFT's may be formulated as a single LFT.*

An example that demonstrates this property is given by the system shown in Figure 3.3. This system is composed of two models, P and Q, that are defined as constant matrices with elements P_{11}, P_{12}, P_{21} and P_{22} for P and similarly Q_{11}, Q_{12}, Q_{21} and Q_{22} for Q. Also, there are operators Δ_P and Δ_Q that are associated with each model through LFT interconnections. Of course, this example could be performed using frequency-varying models but the principle of adding LFT's is more simply presented using constant matrices.

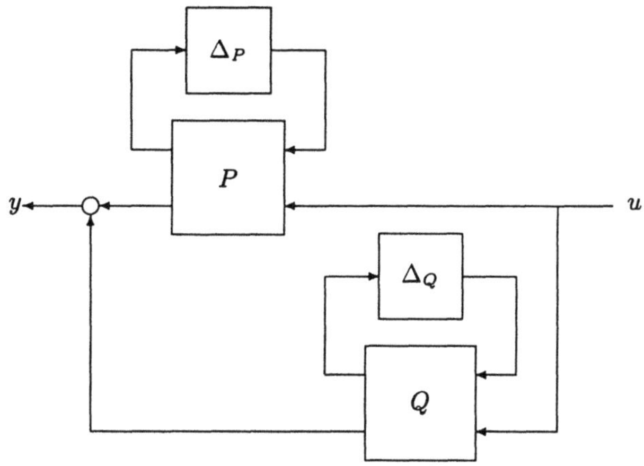

Fig. 3.3: Interconnection of Multiple LFT Models

A single LFT can be defined to represent the interconnection of the multiple LFT's in Figure 3.3 and satisfy Property 3.1.3. Consider the output of a linear fractional transformation with structured operators.

$$
\begin{aligned}
y &= F_u\left(\begin{bmatrix} P_{11} & 0 & P_{12} \\ 0 & Q_{11} & Q_{12} \\ P_{21} & Q_{21} & P_{22}+Q_{22} \end{bmatrix}, \begin{bmatrix} \Delta_P & 0 \\ 0 & \Delta_Q \end{bmatrix}\right) u \\
&= \Bigg(P_{22}+Q_{22} \\
&\quad + \begin{bmatrix} P_{21} & Q_{21} \end{bmatrix} \begin{bmatrix} \Delta_P & 0 \\ 0 & \Delta_Q \end{bmatrix} \left(I - \begin{bmatrix} P_{11} & 0 \\ 0 & Q_{11} \end{bmatrix} \begin{bmatrix} \Delta_P & 0 \\ 0 & \Delta_Q \end{bmatrix}\right)^{-1} \begin{bmatrix} P_{12} \\ Q_{12} \end{bmatrix}\Bigg) u \\
&= \Bigg(P_{22}+Q_{22} \\
&\quad + \begin{bmatrix} P_{21}\Delta_P & Q_{21}\Delta_Q \end{bmatrix} \begin{bmatrix} (I-P_{11}\Delta_P)^{-1} & 0 \\ 0 & (I-Q_{11}\Delta_Q)^{-1} \end{bmatrix} \begin{bmatrix} P_{12} \\ Q_{12} \end{bmatrix}\Bigg) u \\
&= \left(P_{22}+Q_{22}+\begin{bmatrix} P_{21}\Delta_P & Q_{21}\Delta_Q \end{bmatrix} \begin{bmatrix} (I-P_{11}\Delta_P)^{-1} P_{12} \\ (I-Q_{11}\Delta_Q)^{-1} Q_{12} \end{bmatrix}\right) u \\
&= (P_{22}+P_{21}\Delta_P (I-P_{11}\Delta_P)^{-1} P_{12} \\
&\quad + Q_{22}+Q_{21}\Delta_Q (I-Q_{11}\Delta_Q)^{-1} Q_{12}) u \\
&= \left(F_u(P,\Delta_P)+F_u(Q,\Delta_Q)\right) u
\end{aligned}
\tag{3.4}
$$

The equation for y in (3.4) demonstrates that the interconnection of multiple LFT's can indeed be replaced by a single LFT.

$$F_u\left(\begin{bmatrix} P_{11} & 0 & P_{12} \\ 0 & Q_{11} & Q_{12} \\ P_{21} & Q_{21} & P_{22}+Q_{22} \end{bmatrix}, \begin{bmatrix} \Delta_P & 0 \\ 0 & \Delta_Q \end{bmatrix}\right)$$
$$= F_u\left(P, \Delta_P\right) + F_u\left(Q, \Delta_Q\right) \tag{3.5}$$

Stability of multiple linear fractional transformations can be evaluated by analyzing the single equivalent interconnection; however, this evaluation must consider more than the traditional requirement of poles in the left-half plane. LFT's must also be well-posed in order to be considered stable. This well-posedness condition requires all internal signals, even those that are not measured, to be finite and can be determined by applying the small gain theorem of Theorem 2.3.1 to the interconnection.

3.2 Structured Uncertainty

The concept of uncertainty is formulated as a set of norm bounded operators, $\mathbf{\Delta}$, associated with a nominal plant, P, through an LFT feedback relationship. A family of plants, $\mathcal{P}$, arises through consideration of $F_u(P, \Delta)$ for every $\Delta \in \mathbf{\Delta}$. It is assumed the true plant model lies within this family of plants so if every element of the set $\mathcal{P}$ is stable then the true plant is stable. Alternatively stated, if the nominal plant P is robustly stable to the uncertainty $\mathbf{\Delta}$ then the true plant is stable.

Modeling the uncertainty as a norm bounded operator might lead to overly conservative models; however, the uncertainty description can be made more accurate by including frequency information. It is often possible to formulate a model of a physical system that is accurate at low frequencies but is less accurate for representing the system response at high frequencies. A frequency weighted transfer function, W, is generally associated with each uncertainty element to describe magnitude and phase uncertainty as it varies with frequency.

There are several general ways uncertainty can enter a system model in a linear fractional manner. Two typical types of uncertainty are termed multiplicative and additive uncertainty. Multiplicative uncertainty can be either on the input or output of a system. Systems with these types of uncertainty are easily described in block diagram form. Figure 3.4 represents the linear fractional transformation for a plant with input multiplicative uncertainty while the plant with output multiplicative uncertainty is given in Figure 3.5. Additive uncertainty is represented in Figure 3.6.

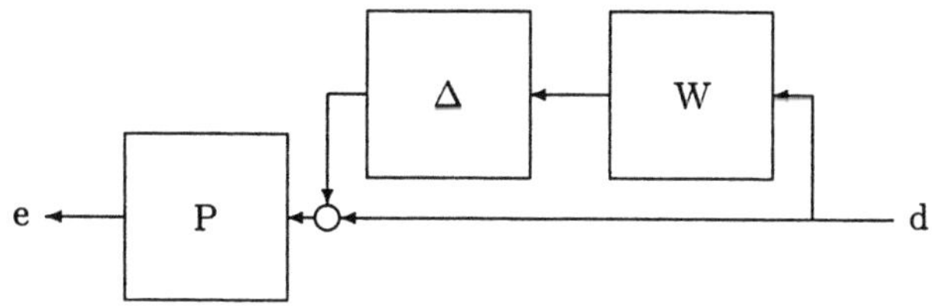

Fig. 3.4: Family of Plants $\mathcal{P} = P(I + \Delta W)$ with Input Multiplicative Uncertainty

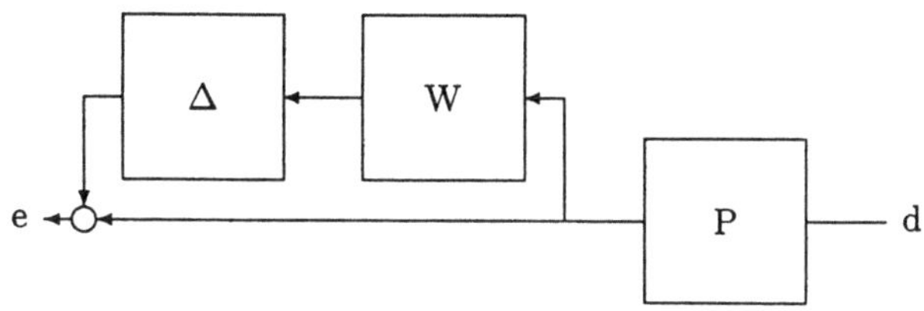

Fig. 3.5: Family of Plants $\mathcal{P} = (I + \Delta W)P$ with Output Multiplicative Uncertainty

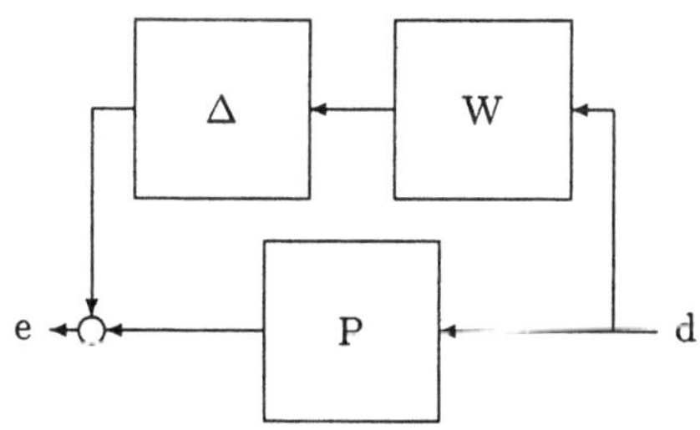

Fig. 3.6: Family of Plants $\mathcal{P} = P + \Delta W$ with Additive Uncertainty

Uncertainty can also be associated with specific elements of the system. These parametric uncertainties are usually associated with a system operator in a feedback relationship. The number of input and output signals of the system operator is increased to account for the additional feedback signals associated with the uncertainty operator. This can be demonstrated by considering the set of plants $\mathcal{P}$ generated by a system with an unknown pole.

$$\mathcal{P} = \left\{ \frac{1}{(s+1)(s+x)} : x \in [2,3] \right\} \tag{3.6}$$

A norm-bounded real scalar uncertainty parameter δ can be introduced to account for the possible variation in pole value. The set of plants can be written in the LFT framework using this uncertainty operator and Definition 3.1.1

$$\mathcal{P} = \left\{ F_u(P,\delta) \ : \ P = \begin{bmatrix} \frac{-1}{s+2.5} & \frac{-1}{s+2.5} \\ \frac{1}{(s+1)(s+2.5)} & \frac{1}{(s+1)(s+2.5)} \end{bmatrix}, \|\delta\|_\infty \le 1, \delta \in \mathbf{R} \right\} \tag{3.7}$$

A complex system with several uncertainty operators can be expressed as an LFT with a single uncertainty operator using Property 3.1.3. This operator is structured as a block diagonal operator with each block associated with the individual uncertainty operators. There are two main types of uncertainty blocks known as full-block and scalar-block uncertainties.

A full-block uncertainty is a matrix with unknown parameters in every element. This type of block is used to describe unstructured uncertainty in a group of signals.

Definition 3.2.1 *A full-block uncertainty $\Delta \in \mathbf{C}^{n \times m}$ has unknown elements Δ_{ij} for every $i \in [1,n]$ and $j \in [1,m]$.*

A repeated scalar-block introduces more structure into the uncertainty description since only the diagonal elements of the matrix contain unknown elements while the remaining elements are zero. Furthermore, the diagonal elements are identical. This type of uncertainty is used to relate input-output signal pairs with the same uncertainty parameter.

Definition 3.2.2 *A repeated scalar-block uncertainty $\Delta \in \mathbf{C}^{n \times n}$ has zero valued elements except an unknown parameter δ along the diagonal such that $\Delta = \delta I_n$. A scalar-block uncertainty is a repeated scalar-block uncertainty of dimension 1.*

The single structured uncertainty block used for robust stability analysis is formally defined in terms of these blocks. Let integers m,n,p define the number of real scalar, complex scalar, and complex full blocks respectively. Define integers $R_1, \ldots, R_m$ such that the ith repeated, scalar block of real parametric uncertainty is of dimension $R_i \times R_i$. Define similar integers $C_1, \ldots, C_n$ to denote the dimension of the complex repeated, scalar blocks. The structured uncertainty description Δ is assumed to be norm bounded and belonging to the sct $\boldsymbol{\Delta}$.

$$\begin{aligned}\boldsymbol{\Delta} = \{\Delta = diag\left(\delta_1^R I_{R_1}, \ldots, \delta_m^R I_{R_m}, \delta_1^C I_{C_1}, \ldots, \delta_n^C I_{C_n}, \Delta_1, \ldots, \Delta_p\right) \\ : \delta_i^R \in \mathbf{R}, \delta_i^C \in \mathbf{C}, \Delta_i \in \mathbf{C}^{c_i \times c_i}\}\end{aligned} \tag{3.8}$$

Real parametric uncertainty is allowed to enter the problem as scalar or repeated scalar blocks while complex uncertainty can enter the problem as scalar, repeated scalar, or full blocks. Complex uncertainty parameters allow uncertainty in magnitude and phase and is thus useful for modeling variations in signals. Conversely, uncertainty in physical parameters in the equations of motion are typically real and constant so their variations are often more accurately modeled as real parametric uncertainty. The robustness analysis will be less conservative by accounting for this structure to accurately describe the model uncertainty.

3.3 Structured Singular Value : μ

The general framework for robust stability analysis is given in Figure 3.7.

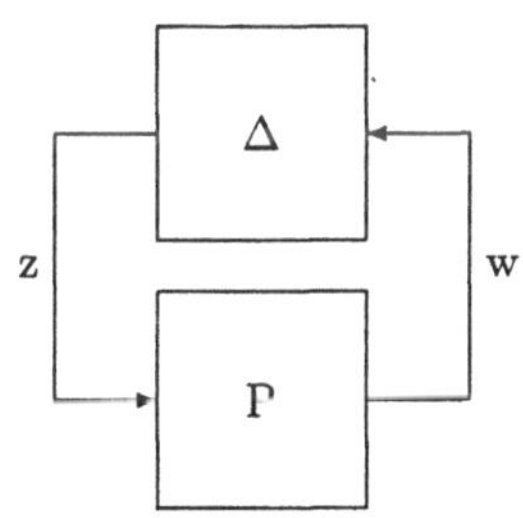

Fig. 3.7: Linear Fractional Transformation System for Robust Stability Analysis using μ

The plant operator $P(s) \in \mathcal{RH}_\infty$ is a stable rational transfer function matrix representing the aeroelastic dynamics. A norm bounded set $\boldsymbol{\Delta} \in \mathcal{RH}_\infty$ is defined such that $\Delta(s) \in \boldsymbol{\Delta}$ describes the modeling errors in P through a feedback relationship.

The robustness of P with respect to the set $\boldsymbol{\Delta}$ can be determined using the small gain theorem as presented in Lemma 2.4.1. This condition guarantees stability for any value $\Delta \in \boldsymbol{\Delta}$ if $||P||_\infty < 1$. This robustness condition may be overly conservative since it does not account for structure in the uncertainty operator. The structured singular value, μ, is defined as an alternative measure of robustness [98].

Definition 3.3.1 *Given the complex transfer function matrix $P \in \mathcal{RH}_\infty$ and associated norm bounded set of uncertainty operators $\mathbf{\Delta}$, define μ.*

$$\mu(P) = \frac{1}{\min_{\Delta \in \mathbf{\Delta}} \{\bar{\sigma}(\Delta) \ : \ \det(I - P\Delta) = 0\}}$$

with $\mu = 0$ if no $\Delta \in \mathbf{\Delta}$ exists such that $\det(I - P\Delta) = 0$.

μ is an exact measure of robust stability for systems with structured uncertainty. The value of μ determines the allowable size of uncertainty matrices for which the plant is robustly stable as demonstrated in Theorem 3.3.2 [98].

Theorem 3.3.2 *Given the system in Figure 3.7, P is robustly stable with respect to the set $\mathbf{\Delta}$ which is norm bounded by real scalar α such that $||\Delta||_\infty \leq \alpha$ for all $\Delta \in \mathbf{\Delta}$ if and only if $\mu(P) < \frac{1}{\alpha}$.*

The model P is usually weighted such that the range of modeling errors is described by the uncertainty set $\mathbf{\Delta}$ which is norm bounded by 1.

$$\mathbf{\Delta} = \{\Delta \ : \ ||\Delta||_\infty \leq 1\} \tag{3.9}$$

Theorem 3.3.3 presents the specific condition for robust stability that will be used for unity norm-bounded uncertainty sets $\mathbf{\Delta}$.

Theorem 3.3.3 *Given the system in Figure 3.7, P is robustly stable with respect to the set $\mathbf{\Delta}$ with $||\Delta||_\infty \leq 1$ for all $\Delta \in \mathbf{\Delta}$ if and only if $\mu(P) < 1$.*

A value of $\mu < 1$ implies no perturbation within the set $\mathbf{\Delta}$ exists that will destabilize the feedback system. This can also be interpreted as saying the true plant dynamics are stable assuming these dynamics lie within the range generated by the nominal model dynamics coupled with the set of perturbations represented by $\mathbf{\Delta}$.

Obviously, μ is dependent on the block structure of the set $\mathbf{\Delta}$. The robust stability properties computed by μ will only be accurate if a realistic uncertainty operator is chosen. μ may be arbitrarily greater when computed with respect to an unstructured uncertainty operator as compared to a high structured uncertainty operator. Definition 3.3.1 demonstrates the μ condition of Theorem 3.3.3 is equivalent to the small gain condition of Lemma 2.4.1 when the uncertainty is unstructured.

Unfortunately, μ is a difficult quantity to compute. Closed form solutions exist to exactly compute μ for only a small number of block structures for $\mathbf{\Delta}$. Upper and lower bounds are used to compute μ for generalized uncertainty block structures. A derivation of the upper bound, which represents a limit on the worst-case robustness properties, is given in Appendix A.3.

A.3 Upper Bound for μ

The structured singular value, μ, is a difficult quantity to compute and appears to be an NP-hard problem [20]. Alternatively, upper and lower bounds have been derived that may be easily computed. The lower bound may be solved by a power iteration which extracts the smallest destabilizing perturbation [98]. The upper bound may be posed as an optimization problem to present a conservative bound on the worst-case stability properties of the system. The analysis of aeroservoelastic dynamics is usually more interested in worst-case stability margins so this Appendix concentrates on the optimization formulation to compute a μ upper bound.

Question A.3.1 *What is an upper bound for μ with respect to a general structured uncertainty description that may be computed?*

A simple upper bound for $\mu(P)$ is the maximum singular value of the plant, $\overline{P}$, as is immediately obvious from Definition 3.3.1.

$$\mu(P) \leq \overline{\sigma}(P) \tag{A.3.1}$$

This upper bound is essentially the robustness as measured by the small gain theorem and may be highly conservative when the uncertainty operator is structured. The conservatism in this upper bound can be reduced by considering two properties of singular values.

Property A.3.2 *The maximum singular value has a multiplicative property.*

$$\overline{\sigma}(P)\overline{\sigma}(P^{-1}) = 1$$

Property A.3.3 *The maximum singular value has a triangle inequality property.*

$$\overline{\sigma}(DP) \leq \overline{\sigma}(D)\overline{\sigma}(P)$$

These properties demonstrate the maximum singular value of a scaled matrix is less than or equal to that of an unscaled matrix.

Lemma A.3.4 *Given matrix $P \in \mathbf{C}^{n \times n}$ and invertible $D \in \mathbf{C}^{n \times n}$, then*

$$\overline{\sigma}(DPD^{-1}) \leq \overline{\sigma}(P)$$

Recall the definition of the uncertainty structure given in Section 3.2.

$$\mathbf{\Delta} = \left\{ \Delta = diag\left(\delta_1^R I_{R_1}, \ldots, \delta_m^R I_{R_m}, \delta_1^C I_{C_1}, \ldots, \delta_n^C I_{C_n}, \Delta_1, \ldots, \Delta_p\right) \right. \\ \left. : \delta_i^R \in \mathbf{R}, \delta_i^C \in \mathbf{C}, \Delta_i \in \mathbf{C}^{c_i \times c_i} \right\}$$

A set of scaling matrices are defined with structure similar to the uncertainty structure. The set $\mathcal{D}$ contains blocks to scale with every complex and real uncertainty block [98].

$$\mathcal{D} = \{D = diag\left(D_1^R, \ldots, D_m^R, D_1^C, \ldots, D_n^C, d_1^c I_{c_1}, \ldots, d_p^c I_{c_p}\right) \\ : D_i^R \in \mathbf{C}^{R_i \times R_i}, D_i^C \in \mathbf{C}^{C_i \times C_i}, d_i \in \mathbf{C}, D = D^* > 0\}$$

The set of scalings $\mathcal{G}$ affect only the real parametric uncertainty blocks [39].

$$\mathcal{G} = \{G = diag\left(G_1, \ldots, G_m, 0, \ldots, 0\right) : G_i \in \mathbf{C}^{R_i \times R_i}, G = G^*\}$$

There are several advantageous properties of these scalings.

Property A.3.5 *Symmetric factors of scalings are valid scalings.*

$$D^{-1}, D^{-\frac{1}{2}}, D^{\frac{1}{2}} \in \mathcal{D} \text{ for every } D \in \mathcal{D}$$

Property A.3.6 *Every scaling $D \in \mathcal{D}$ and $G \in \mathcal{G}$ commutes with every uncertainty $\Delta \in \mathbf{\Delta}$.*

$$D\Delta = \Delta D$$
$$G\Delta = \Delta G$$

μ is shown to be invariant under transformations utilizing the scaling matrices $\mathcal{D}$ [98]. This invariance arises as a direct consequence of Properties A.3.5 and A.3.6 which demonstrate $\mathbf{\Delta} = \mathcal{D}\mathbf{\Delta}\mathcal{D}^{-1}$. Thus, computing μ with respect to $\mathbf{\Delta}$ is equivalent to computing μ with respect to $D\mathbf{\Delta}D^{-1}$ for all $D \in \mathcal{D}$. A new upper bound for μ is formulated using these scaling matrices.

$$\mu(P) \leq \inf_{D \in \mathcal{D}} \overline{\sigma}\left(DPD^{-1}\right) \leq \overline{\sigma}(P) \tag{A.3.2}$$

This upper bound is less conservative than the unscaled upper bound due to the final inequality which is a consequence of Lemma A.3.4. This upper bound uses a search over all valid scaling matrices to find the optimal scaling matrix which reduces the conservatism in the robustness computation. This upper bound can be formulated as a maximum eigenvalue condition as demonstrated in Lemma A.3.7.

Lemma A.3.7 *The constant matrix P is robustly stable and $\mu(P) < 1$ with respect to the set $\mathbf{\Delta}$ if*

$$\inf_{D \in \mathcal{D}} \|D^{\frac{1}{2}} P D^{-\frac{1}{2}}\|_\infty < 1$$

or equivalently

$$\inf_{D \in \mathcal{D}} \overline{\lambda}\left(P^* D P - D\right) < 0$$

This upper bound allows scalar uncertainty parameters to lie within a norm bounded disk centered in the complex plane which may be conservative if these parameters are real. The conservatism can be reduced by introducing the second set of scaling matrices $\mathcal{G}$ which affect only the real parametric uncertainty blocks. These scalings restrict the uncertainty set to lie within a norm bounded disk covering the real axis between 1 and -1, but that disk is allowed to be centered off the origin. A new upper bound is formulated with both $\mathcal{D}$ and $\mathcal{G}$ scalings [39, 141].

$$\mu(P) \leq \sqrt{max(0, \alpha_*)} \tag{A.3.3}$$

where α_* is defined as

$$\alpha_* = \inf_{\substack{D\in\mathcal{D}\\ G\in\mathcal{G}}} \min_{\alpha\in\mathbf{R}} \{\alpha \; : \; P^*DP + \jmath(GP - P^*G) - \alpha D \leq 0\}$$

This upper bound is presented as an optimization in Lemma A.3.8.

Lemma A.3.8 *The constant matrix P is robustly stable and $\mu(P) < 1$ with respect to the set $\mathbf{\Delta}$ if*

$$\inf_{\substack{D\in\mathcal{D}\\ G\in\mathcal{G}}} \{\alpha \; : \; \overline{\lambda}(P^*DP + \jmath(GP - P^*G) - \alpha D) \leq 0\} < 1$$

The $\mathcal{G}$ matrices are zero if there are no real parameter uncertainty blocks in the system so the condition in Lemma A.3.8 reverts to the condition in Lemma A.3.7. When the $\mathcal{G}$ matrices are nonzero the additional terms in the new upper bound condition arise by using an additional constraint on the signals in the system [39]. Consider some signal in the closed loop system relating some uncertainty signal. The system equation, ignoring the extraneous error and disturbance signals, can be simply written.

$$\Delta Px = x \tag{A.3.4}$$

Allow the uncertainty to be a single repeated real parameter, $\Delta = \delta I$ with $\delta \in \mathbf{R}$, and take the complex conjugate transpose of the entire equation.

$$\delta x^* P^* = x^* \tag{A.3.5}$$

Combining this formula with a scaling $G \in \mathcal{G}$ leads to the following.

$$x^*GPx = \delta x^*P^*GPx = x^*P^*G(\delta Px) = x^*P^*Gx \tag{A.3.6}$$

The following matrix is derived from the right and left sides of this equation.

$$x^*(GP - P^*G)x = 0 \tag{A.3.7}$$

Consider the matrix term of the complex uncertainty robustness upper bound scaled by x.

$$x^*(P^*DP - \alpha D)x < 0 \tag{A.3.8}$$

Combining the two equations leads to the new upper bound condition.

$$x^*\left(P^*DP + \jmath(GP - P^*G) - \alpha D\right)x < 0 \tag{A.3.9}$$

This μ upper bound is directly related to robust stability analysis resulting from the small gain theorem [39]. The $\mathcal{D}$ scaling matrices utilize structure in the uncertainty while the $\mathcal{G}$ matrices utilize additional phase information provided by real parametric uncertainty blocks [141]. This robustness condition is also related to a stability criterion used in absolute stability theory where the scaling matrices enter as Popov multipliers [59].

The μ upper bound conditions in Lemma A.3.7 and A.3.8 compute the robustness of a constant matrix to simplify the equations without carrying ω terms throughout the derivations. In practice the robustness must be computed for physical systems described by linear, time-invariant transfer functions, $G \in \mathcal{RH}_\infty$, which are complex valued matrix functions of frequency. The upper bound conditions are computed for these transfer functions by including an additional search over frequency to find the worst-case upper bound.

Lemma A.3.9 *The transfer function matrix $P(\jmath\omega) \in \mathcal{RH}_\infty$ is robustly stable and $\mu(P) < 1$ with respect to the set $\mathbf{\Delta}$ if*

$$\max_{\omega} \inf_{\substack{D \in \mathcal{D} \\ G \in \mathcal{G}}} \left\{\alpha : \overline{\lambda}\left(P(\jmath\omega)^*DP(\jmath\omega) + \jmath(GP(\jmath\omega) - P^*(\jmath\omega)G) - \alpha D\right) \leq 0\right\} < 1$$

Established software packages compute the μ upper bound by finding optimal scaling matrices at distinct frequency points [5]. Frequency-varying scalings commute with linear time-invariant (LTI) uncertainty which may not accurately capture the nature of the modeling errors. Restricting the scaling matrices to be constant with no frequency variation allows consideration of uncertainty operators with linear time-varying (LTV) behavior. The μ upper bound for complex uncertainty is a necessary and sufficient condition for robust stability of an LTI plant with complex LTV uncertainty for any block structure [117]. Computing robustness measures for LTI systems with complex LTV uncertainty is addressed for finite energy signals in Reference [3] and for finite gain signals in Reference [61]. Related papers have studied robustness conditions for systems with some type of rate bounds on the LTV uncertainty [108].

In general a system will require both LTI and LTV uncertainties blocks to accurately describe the modeling errors. Parametric uncertainties are associated with physical elements of the system which are necessarily real and often time-varying. Dynamic uncertainties are used to introduce magnitude and phase variations in uncertain signals and should be considered time-invariant. Introduce two sets of structured uncertainty operators.

$$\begin{aligned}\mathbf{\Delta}_{TV}(t) = \{\Delta(t) = diag\,(\delta_1^R I_{R_1}, \ldots, \delta_m^R I_{R_m}, \delta_1^C I_{C_1}, \ldots, \delta_n^C I_{C_n}, \Delta_1, \ldots, \Delta_p) \\ : \delta_i^R(t) \in \mathbf{R}, \delta_i^C(t) \in \mathbf{C}, \Delta_i(t) \in \mathbf{C}^{c_i \times c_i}\}\end{aligned}$$

$$\begin{aligned}\mathbf{\Delta}_{TI}(\omega) = \{\Delta(\omega) = diag\,(\delta_1^R I_{R_1}, \ldots, \delta_m^R I_{R_m}, \delta_1^C I_{C_1}, \ldots, \delta_n^C I_{C_n}, \Delta_1, \ldots, \Delta_p) \\ : \delta_i^R(\omega) \in \mathbf{R}, \delta_i^C(\omega) \in \mathbf{C}, \Delta_i(\omega) \in \mathbf{C}^{c_i \times c_i}\}\end{aligned}$$

The set $\mathbf{\Delta}_{TV}$ contains all linear time-varying uncertainty operators that are allowed to change infinitely fast as a function of time. Other researchers have investigated robustness with respect to uncertainty operators with known rate variations; however, the formulation here will assume infinitely-fast variations are possible [108]. Correspondingly, the set $\mathbf{\Delta}_{TI}$ contains all linear time-invariant uncertainty operators that are allowed to vary with frequency. These sets are allowed to enter the problem as in Figure A.3.1.

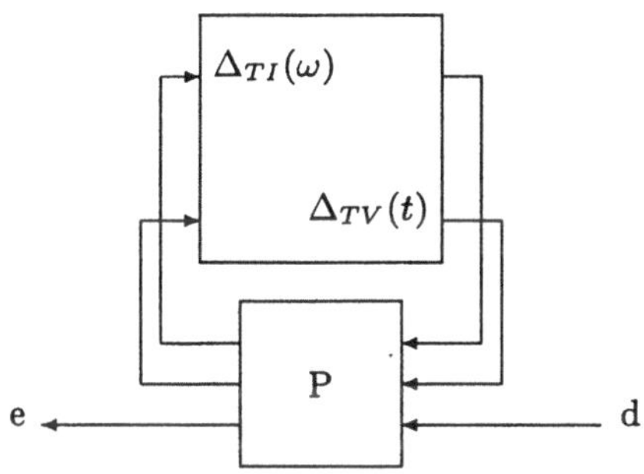

Fig. A.3.1: Block Diagram with LTI and LTV Uncertainty

Associated with these uncertainty descriptions are scaling matrices. Constant scaling matrices, $\mathcal{D}_{TV}$, are structured like $\mathcal{D}$ to associate with the time-varying uncertainty, $\mathbf{\Delta}_{TV}$. Frequency dependent scalings, $\mathcal{D}_{TI}(\omega)$, are associated with the time-invariant uncertainty, $\mathbf{\Delta}_{TI}$. The block structure of each of these sets matches the block structure of the corresponding uncertainty set.

The uncertainty and scaling matrix sets may be combined into single sets for ease of notation.

$$\begin{aligned}\overline{\mathbf{\Delta}} &= diag\,\{\Delta_{TI}(\omega), \Delta_{TV}(t)\} \\ \overline{\mathcal{D}} &= diag\,\{D_{TI}(\omega), D_{TV}\} \\ \overline{\mathcal{G}} &= diag\,\{G_{TI}(\omega), G_{TV}\}\end{aligned}$$

Lemma A.3.10 describes the robustness upper bound condition.

Lemma A.3.10 *Given some $\alpha > 0$ and norm bounded uncertainty $\overline{\boldsymbol{\Delta}}$ such that*

$$\max_{\omega} \bar{\sigma}\left(\Delta_{TI}(\omega)\right) \leq \frac{1}{\alpha}$$

and

$$\max_{t} \bar{\sigma}\left(\Delta_{TV}(t)\right) \leq \frac{1}{\alpha}$$

then the system described by transfer function $P(\jmath\omega)$ is robustly stable with respect to $\overline{\boldsymbol{\Delta}}$ if there exist scaling matrices $D \in \overline{\mathcal{D}}$ and $G \in \overline{\mathcal{G}}$ such that:

$$\max_{\omega} \overline{\lambda}\left(P(\jmath\omega)^* D(\omega) P(\jmath\omega) + \jmath(G(\omega)P(\jmath\omega) - P(\jmath\omega)^* G(\omega)) - \alpha^2 D(\omega)\right) < 0$$

The search for optimal scaling matrices that achieve the lowest robustness value can be formulated as an optimization to minimize α subject to a constraint function associated with an eigenvalue condition. An additional constraint arises from restricting the $\mathcal{D}$ scaling matrices to be positive definite.

$$\begin{aligned} & \min_{\omega} \alpha && \text{(A.3.10)} \\ \text{subject to} \quad & -M^*DM - \jmath(GM - M^*G) + \alpha^2 D > 0 \\ & D > 0 \end{aligned}$$

This upper bound condition is written in this format to ease understanding of the function and develop solution algorithms. Specifically, the minimization problem as presented above with the eigenvalue constraints is a linear matrix inequality (LMI). The LMI is a convenient framework to utilize since there are common properties global to all LMI problems [19]. The main property states an LMI is a convex function of the variables which allows solutions to be computed via standard convex optimization algorithms. The Ellipsoid Method and the interior point algorithm of the Method of Centers are shown to be efficient for LMI solutions [15, 18, 45].

It is important to note this optimization involves a search over frequency. Only a finite set of frequency points are used in practice and engineering judgment must be used in choosing the set of frequency points. Testing the robustness condition at more frequency points will increase the accuracy of the upper bound, but the number of free parameters will rise accordingly and increase computational time [73].

CHAPTER 4

AEROSERVOELASTICITY

4.1 Lagrangian Derivation of Equations of Motion

The equations of motion of an aircraft or a flight vehicle can be derived by considering various forces acting on the system. In classical mechanics, a Lagrangian definition of the equations results in an elegant form of deriving the equilibrium state of a flying vehicle. This requires certain physical concepts leading to the definition of the energy and the kinematics. The energy can be expressed in terms of kinetic energy, potential energy, and the virtual work done. The kinematics are generally expressed with respect to an inertial coordinate system which is assumed to be at rest. In the case of an Earth-bound vehicle, the inertial system is assumed to be fixed to the center of the Earth where it is assumed that the angular velocities and the linear motions of the Earth have little effect on the dynamics of the aircraft.

Let the position vector of a mass point on the aircraft be described by

$$\boldsymbol{r}_i = \boldsymbol{r}_r + \boldsymbol{r}_{b_i} + \boldsymbol{q}_{e_i} \tag{4.1}$$

where

$\boldsymbol{r}_i$	is the position vector of the mass point i
$\boldsymbol{r}_r$	is the position vector of the rigid body-fixed coordinate system origin O with respect to inertial space
$\boldsymbol{r}_{b_i}$	is the position vector of mass point i with respect to origin O
$\boldsymbol{q}_{e_i}$	is a vector of the deformation of the mass i with respect to the rigid body-fixed coordinate system

If p, q, and r are the angular velocities of the aircraft measured about the rigid body-fixed axis system, then the velocity of the mass can be written as

$$\dot{\boldsymbol{r}}_i = \frac{\partial}{\partial t}(\boldsymbol{r}_i) + \boldsymbol{\omega} \times \boldsymbol{r}_i \tag{4.2}$$

where $\boldsymbol{\omega} = p\hat{\boldsymbol{i}} + q\hat{\boldsymbol{j}} + r\hat{\boldsymbol{k}}$ in terms of unit direction vectors $\hat{\boldsymbol{i}}, \hat{\boldsymbol{j}}, \hat{\boldsymbol{k}}$. Equations (4.1) and (4.2) in matrix notation for all mass points can be written as

$$q = \Phi_r q_r + q_e + \Phi_\delta q_\delta \tag{4.3}$$

$$\dot{q} = \Phi_r \dot{q}_r + \dot{q}_e + \Phi_\delta \dot{q}_\delta \Omega(\Phi_r q_r + q_e + \Phi_\delta q_\delta) \tag{4.4}$$

for

$$\Omega = \begin{bmatrix} \ddots & & 0 \\ & \tilde{\omega} & \\ 0 & & \ddots \end{bmatrix} \tag{4.5}$$

and

$$\tilde{\omega} = \begin{bmatrix} 0 & -R & Q \\ R & 0 & -P \\ -Q & P & 0 \end{bmatrix} \tag{4.6}$$

where $\{q_r, q_\delta\}$ and $\{\dot{q}_r, \dot{q}_\delta\}$ are rigid body and control surface displacement and velocity vectors, $\{\Phi_r, \Phi_\delta\}$ are rectangular matrices of the rigid body and control mode shapes, and q_e are the elastic coordinates.

Derivation of the equations of motion based on energy conservation begins with the Lagrangian principle

$$\frac{d}{dt}\left(\frac{\partial L}{\partial \dot{\eta}}\right) - \frac{\partial L}{\partial \eta} = 0 \tag{4.7}$$

where

$$L = KE - PE - VW \tag{4.8}$$

in which KE is the kinetic energy, PE is the potential energy, and VW is the virtual work.

To simplify the algebra in the following development, the displacement vector, q, and the corresponding velocity, $\dot{q}$, will be expressed in the generalized coordinate system, that is

$$q = \left[\Phi_r \; \Phi_e \; \Phi_\delta\right] \begin{pmatrix} \eta_r \\ \eta_e \\ \eta_\delta \end{pmatrix} = \Phi\eta \tag{4.9}$$

where $\{\Phi_e, \Phi_\delta\}$ are rectangular matrices of the vibration and control modes. The velocity vector in terms of the generalized coordinates is given by

$$\dot{q} = \Phi\dot{\eta} + \Omega\Phi\eta. \tag{4.10}$$

Assuming that the mass matrix, M, of the aircraft system is known, the kinetic energy of the system is given by

$$KE = \frac{1}{2}\dot{q}M\dot{q}. \tag{4.11}$$

Substituting (4.10) into (4.11) gives

$$KE = \frac{1}{2}\left[\Phi\dot{\eta} + \Omega\Phi\eta\right]^T M \left[\Phi\dot{\eta} + \Omega\Phi\eta\right]. \tag{4.12}$$

If the stiffness matrix of the aircraft system is denoted by K, the potential energy of the system is

$$PE = \frac{1}{2}q^T K q = \frac{1}{2}\eta^T \Phi^T K \Phi \eta. \tag{4.13}$$

The external forces acting on the system can be classified as those dependent on the elastic deformation and the oscillatory motion of the aircraft, and those independent of the motion. The motion dependent forces are

- rigid body aerodynamic forces
- unsteady aerodynamic forces
- body forces

and motion independent forces are

- atmospheric gust loads
- engine thrust

The computed airloads in a rigid configuration involves several parameters such as angle of attack α, sideslip angle β, angular velocities p, q, and r, and the control deflections δ. Since the aerodynamic forces are nonlinear functions of these parameters, a set of panel loads are computed at discrete values of these parameters and stored in a matrix array. In the analysis the air load vector for derived values of the parameters are computed by interpolation. Hence the rigid air load vector is represented by

$$F_{\text{rigid}} = F\left(\alpha, \beta, P, Q, R, \delta\right). \tag{4.14}$$

The virtual work done by this force is

$$VW_1 = \eta^T \Phi^T F_{\text{rigid}}. \tag{4.15}$$

Aerodynamic forces arising from elastic deformation and oscillatory motion can be given by

$$F_{\text{flex}} = \bar{q} A_e(k) q \tag{4.16}$$

where $A_e(k)$ is an influence coefficient matrix and $\bar{q}$ is the dynamic pressure. The contribution to the virtual work from the unsteady forces is

$$VW_2 = \bar{q}\eta^T \Phi^T A_e(k) \Phi \eta \tag{4.17}$$

$$= \bar{q}\eta^T \left[Q(k)\right] \eta \tag{4.18}$$

where $Q(k)$ are the generalized forces due to $\{\eta_r, \eta_e, \eta_\delta\}$ coordinates. As $k \to 0$, $Q(k)$ corresponds to the steady aerodynamic loads. However the airloads due to the rigid body modes are already included in the F_{rigid} vector. Therefore the steady air load components due to η_r should be subtracted from $Q(k)$ to give

$$\overline{Q}(k) = Q(k) - \left[\overbrace{Q(0)}^{\eta_r} \ \overbrace{0}^{\eta_e}\right] \tag{4.19}$$

and the virtual work is

$$VW_2 = \bar{q}\eta^T \overline{Q}(k)\eta. \tag{4.20}$$

Let ϕ, θ, ψ be the Euler angles defining the orientation of the body-fixed axis with respect to the Earth. Then the gravity vector with acceleration $\bar{g}$ is

$$\bar{\boldsymbol{g}} = \bar{g} \begin{pmatrix} -\sin\theta \\ \cos\theta \sin\phi \\ \cos\theta \cos\phi \end{pmatrix} \tag{4.21}$$

and the body forces are given by

$$F_{\text{body}} = ME\bar{\boldsymbol{g}} \tag{4.22}$$

where $E = \begin{bmatrix} I_3 \\ I_3 \\ \vdots \end{bmatrix}$, so the virtual work due to body forces is

$$VW_3 = \eta^T \Phi^T ME\bar{\boldsymbol{g}}. \tag{4.23}$$

The dynamic loads in the gust environment can be computed based on the unsteady aerodynamic influence coefficients. The contribution to the virtual work from the gust loads is

$$VW_4 = \eta^T \Phi^T A_e(k)\alpha(\omega)G_{\text{gust}} \tag{4.24}$$

where $\alpha(\omega)$ is an equivalent downwash computed from the gust environment and G_{gust} is a measure of the gust intensity.

Modern tactical fighters are expected to employ thrust vectoring for superior maneuverability. Therefore the work done in a multi-axis vectoring system is an important factor in the analysis of stability and maneuver performance. The virtual work contributed by the engine thrust is

$$VW_5 = \eta^T \Phi^T T \tag{4.25}$$

where T is a vector representing the components of the thrust at specified points.

Having defined the expressions for the energy, it is then permissible to derive the equations of motion using the Lagrangian principle. KE is the kinetic energy given by (4.12), PE is the potential energy given by (4.13), and VW is the virtual work as the sum of the terms given by (4.15), (4.20), (4.23), (4.24), and (4.25). Performing the indicated differentiation with respect to each component of the generalized coordinates (η) and simplifying, the following equations of motion are obtained.

$$\Phi^T \left[\bar{q}F_{\text{rigid}} - \underbrace{ME\bar{g}}_{\text{body-force}} + \underbrace{F_T}_{\text{thrust}} - \underbrace{\bar{q}Q_G G_{\text{gust}}}_{\text{gust}} \right] =$$

$$\left[\underbrace{\Phi^T M \Phi}_{\text{gen-mass}} \right] \begin{pmatrix} \ddot{\eta}_r \\ \ddot{\eta}_e \\ \ddot{\eta}_\delta \end{pmatrix} + \left[+\Phi^T \left(\underbrace{C}_{\text{damping}} + \underbrace{M\Omega + \Omega^T M}_{\text{centrifugal}} \right) \Phi \right] \begin{pmatrix} \dot{\eta}_r \\ \dot{\eta}_e \\ \dot{\eta}_\delta \end{pmatrix}$$

$$+ \left[\underbrace{\Phi^T K \Phi}_{\text{gen-stiffness}} + \underbrace{\Phi^T \Omega^T M \Omega \Phi}_{\text{inertial}} + \bar{q} \underbrace{\overline{Q}(k)}_{\text{unsteady}} \right] \begin{pmatrix} \eta_r \\ \eta_e \\ \eta_\delta \end{pmatrix} . \qquad (4.26)$$

This equation has the same form as the generalized equation of motion for describing structural dynamics with an external forcing function $P(t)$

$$P(t) = M\ddot{\eta} + C\dot{\eta} + K\eta + \bar{q}Q(k)\eta \qquad (4.27)$$

yet also includes the additional terms to perform a nonlinear maneuver analysis. Structural damping C is included with the velocity components. The inertial stiffness matrix given by $\Phi^T \Omega^T M \Omega \Phi$ is of significant importance for the design of the feedback control system because this term contributes to the nonlinear stiffness property of the control surfaces.

The centrifugal inertia given by $\Phi^T \left[M\Omega + \Omega^T M \right] \Phi$ is a function of angular velocities p, q, and r, and gives rise to the load factor imposed on the system. The equation of motion defined by (4.26) can be reduced to a first-order differential equation and solved for the state velocities. However, for performing a maneuver analysis, the velocity and acceleration terms $\dot{\eta}_e$ and $\ddot{\eta}_e$ of the elastic modes may be assumed to be small compared to the corresponding rigid body terms. Then the second set of equations in (4.26) can be solved for η_e and substituted in the first set of equations resulting in the following equations of motion in the rigid body coordinates.

$$M\ddot{\eta}_r = \Xi \left[\bar{q}F_{\text{rigid}} - ME\bar{g} + F_T - \left(M\Omega + \Omega^T M \right) \Phi_r \dot{\eta}_r \right] \qquad (4.28)$$

where Ξ is the aeroelastic correction matrix

$$\Xi = \left[\Phi_r^T + \bar{q}Q_{re}B^{-1}\Phi_e^T \right] \qquad (4.29)$$

in which

$$B = \overline{K}_{ee} - \bar{q}Q_{ee}(k=0). \qquad (4.30)$$

The stiffness matrix $\overline{K}_{ee}$ is an effective stiffness including the centrifugal effects. The expression given by (4.30) can be treated as an eigenvalue problem for the determination of the wing divergence speed ($\bar{q}_{div}$).

The rigid body velocity and acceleration vectors can be written as

$$\dot{\eta}_r = \begin{pmatrix} U \\ V \\ W \\ P \\ Q \\ R \end{pmatrix} \tag{4.31}$$

and

$$\ddot{\eta}_r = \begin{pmatrix} \dot{U} \\ \dot{V} \\ \dot{W} \\ \dot{P} \\ \dot{Q} \\ \dot{R} \end{pmatrix} \tag{4.32}$$

Thus (4.28) reduces to a first-order equation in the state velocities. For a given time history of the control variables, a transient maneuver performance of the aircraft can be determined. If the inertia term $M\ddot{\eta}_r$ is set to zero, (4.28) reduces to a nonlinear steady trim equation.

Transformation of the state equations from the inertial frame of reference to a body-fixed coordinate system for stability and control analysis is given in the Appendix. Extensive theoretical development and solution methods for the aeroelastic equations are covered in the excellent texts [12] and [34]

4.2 Aerodynamic Derivative Representation

Linear unsteady aerodynamics are generally calculated using a panel method for aeroelastic analysis and aeroservoelastic models. Aerodynamic paneling is used to compute the unsteady aerodynamic forces from purely oscillatory motion of each mode for a specified range of frequencies [4, 14, 48, 54]. These aerodynamic forces are tabulated complex values which are only valid for undamped oscillatory motion. They are extended to the Laplace domain using approximate analytic continuation in order to represent aerodynamic damping.

Reference [128] outlines some methods of approximating generalized force coefficients with matrix formulations of rational functions in s, the Laplace variable. These are referred to as Padé approximates in the reduced frequency domain, and are estimates of the time lags associated with unsteadiness in the flow. Reduced frequency k is defined as the ratio of the velocity of vibration to the velocity of the air, and is expressed as $\frac{\omega\bar{c}}{2V}$, for vibration frequency ω, true velocity V, and characteristic length of the half chord $\frac{\bar{c}}{2}$.

A more recent method [138] avoids the aerodynamic lag terms in the state space aeroelastic model for control design and computational efficiency in a dynamic simulation. A linear time-invariant aeroelastic state-space realization for each flight condition is formulated. Nonlinear quasi-steady dynamics are then coupled with the linear aeroelastic equations to create a unified set of simulation equations applicable for general aeroservoelastic analyses.

Justification for the application of forced oscillation linear theory is based on the assumption that the structural vibrations about a fixed mean angle of attack are relatively small, so the relation between system displacements and motion-dependent structural and aerodynamic forces is linear. Generalized forces at extremely low reduced frequency define the aerodynamic stiffness and damping matrices from which aircraft stability and control derivatives can be evaluated analytically. Measured steady aerodynamic data may be substituted in preference to the theoretically derived force coefficients. This helps to impose the constraint that the aerodynamic forces near zero reduced frequency correlate to the measured data.

Methods to determine correction factors to improve analytical calculations of steady and unsteady pressures and forces using high quality aerodynamic data from wind tunnel, flight, or computational fluid dynamics codes are explained in [136]. These correction factors are multipliers applied to aerodynamic downwashes or pressures in aerodynamic calculations to achieve a specified objective. Approaches include matching surface pressure distributions, matching airfoil section properties, and matching total forces, moments, or control derivatives. Extension of these corrections to unsteady transonic aerodynamics was successfully done in [102] to account for nonlinear aeroelastic effects. These methods could also improve general six degree of freedom simulation [27] dynamics that combine quasi-steady and unsteady dynamics for stability and control investigations.

An important consideration for accurate aeroservoelastic modeling is the effect of different reference axes between the equations of motion from aeroelasticity and those from the stability and control discipline. Aircraft stability and control derivatives are defined with respect to a body-fixed axis, yet generalized forces are computed in an inertial reference frame. Inertial and body axis coordinate systems are aligned with each other at trim α_1. Body-fixed axes are either aligned longitudinally along the fuselage (body axis) or oriented into the relative wind (stability axis) at trim angle of attack α_1 with respect to the body. Initial alignment at α_1 does not alter the body-fixed nature of the axis system [90], but a transformation in terms of trim angle of attack α_1 is required to incorporate measured derivative values into the generalized aerodynamic stiffness and damping matrices.

The relationship between generalized force coefficients in the inertial axis system and aircraft *body axis* nondimensional stability derivatives is summarized in Tables 4.1 through 4.4. Longitudinal (symmetric mode) coefficients are represented in Tables 4.1 and 4.2 for the generalized stiffness and damping matrices, respectively. Column headings designate rigid body deflection degrees of freedom. Table entries represent linear or angular force coefficients in terms of force per unit deflection of the particular degree of freedom.

Lateral-directional (antisymmetric mode) ***body axis*** coefficients are similarly incorporated in Tables 4.3 and 4.4.

Table 4.1: Longitudinal Body-fixed Axis Derivative Representation in the Generalized Aerodynamic Stiffness Matrix

	x	z
x	0	0
z	0	0
θ	0	0
	θ	δ
x	$\bar{S}\left(C_{D_\alpha} - \alpha_1 C_{D_u} - \alpha_1 C_{L_{\alpha_1}} + \alpha_1^2 C_{L_u}\right)$	$\bar{S}\left(C_{D_\delta} - \alpha_1 C_{L_\delta}\right)$
z	$\bar{S}\left(C_{L_\alpha} - \alpha_1 C_{L_u} + \alpha_1 C_{D_\alpha} - \alpha_1^2 C_{D_u}\right)$	$\bar{S}\left(C_{L_\delta} + \alpha_1 C_{D_\delta}\right)$
θ	$\bar{S}\bar{c}\left(-C_{m_\alpha} + \alpha_1 C_{m_u}\right)$	$-\bar{S}\bar{c}C_{m_\delta}$

Table 4.2: Longitudinal Body-fixed Axis Derivative Representation in the Generalized Aerodynamic Damping Matrix

	x	z
x	$\frac{\bar{S}}{U_1}\left(C_{D_u} - \alpha_1 C_{L_u}\right)$	$\frac{\bar{S}}{U_1}\left(C_{D_\alpha} - \alpha_1 C_{L_\alpha} - C_L\right)$
z	$\frac{\bar{S}}{U_1}\left(C_{L_u} + \alpha_1 C_{D_u}\right)$	$\frac{\bar{S}}{U_1}\left(C_{L_\alpha} + \alpha_1 C_{D_\alpha}\right)$
θ	$-\frac{\bar{S}}{U_1}\bar{c}C_{m_u}$	$-\frac{\bar{S}}{U_1}\bar{c}C_{m_\alpha}$
	θ	δ
x	$\frac{\bar{S}\bar{c}}{2U_1}\left(C_{D_q} + C_{D_{\dot{\alpha}}} - \alpha_1 C_{L_q} - \alpha_1 C_{L_{\dot{\alpha}}}\right)$	0
z	$\frac{\bar{S}\bar{c}}{2U_1}\left(C_{L_q} + C_{L_{\dot{\alpha}}} + \alpha_1 C_{D_q} + \alpha_1 C_{D_{\dot{\alpha}}}\right)$	0
θ	$\frac{\bar{S}\bar{c}}{2U_1}\bar{c}\left(-C_{m_q} - C_{m_{\dot{\alpha}}}\right)$	0

Table 4.3: Lateral-directional Body Axis Derivative Representation in the Generalized Aerodynamic Stiffness Matrix

	y	ϕ	ψ	δ
y	0	$\bar{S}\left(-C_L - \alpha_1 C_{Y_\beta}\right)\alpha_1$	$\bar{S}C_{Y_\beta}$	$-\bar{S}C_{Y_\delta}$
ϕ	0	$-\bar{S}bC_{l_\beta}\alpha_1$	$\bar{S}bC_{l_\beta}$	$-\bar{S}bC_{l_\delta}$
ψ	0	$-\bar{S}bC_{n_\beta}\alpha_1$	$\bar{S}bC_{n_\beta}$	$-\bar{S}bC_{n_\delta}$

An implicit assumption in Tables 4.1 through 4.4 is that the derivatives are defined with reference to one of the body-fixed axes, that is, the body axis. If defined with respect to another body-fixed reference, the ***stability axis***, longitudinal matrices relating the derivatives to the inertial frame of reference, Tables 4.1 and 4.2, are not affected. However, commonly the lateral-

Table 4.4: Lateral-directional Body Axis Derivative Representation in the Generalized Aerodynamic Damping Matrix

	y	ϕ	ψ	δ
y	$-\frac{\bar{S}}{U_1}C_{Y_\beta}$	$\frac{\bar{S}b}{2U_1}\left(-C_{Y_p} - \alpha_1 C_{Y_{\dot\beta}}\right)$	$\frac{\bar{S}b}{2U_1}\left(-C_{Y_r} + C_{Y_{\dot\beta}}\right)$	0
ϕ	$-\frac{\bar{S}}{U_1}bC_{l_\beta}$	$\frac{\bar{S}b}{2U_1}b\left(-C_{l_p} - C_{l_{\dot\beta}}\alpha_1\right)$	$\frac{\bar{S}b}{2U_1}b\left(-C_{l_r} + C_{l_{\dot\beta}}\right)$	0
ψ	$-\frac{\bar{S}}{U_1}bC_{n_\beta}$	$\frac{\bar{S}b}{2U_1}b\left(-C_{n_p} - C_{n_{\dot\beta}}\alpha_1\right)$	$\frac{\bar{S}b}{2U_1}b\left(-C_{n_r} + C_{n_{\dot\beta}}\right)$	0

directional derivatives are defined with respect to the stability axis, and the lateral-directional derivative representation relative to the inertial frame of reference, Table 4.3 and 4.4, must be modified. In Tables 4.5 and 4.6 the lateral-directional derivatives referenced to the stability axis are properly incorporated into the generalized aerodynamic stiffness and damping matrices, respectively.

Table 4.5: Lateral-directional Stability Axis Derivative Representation in the Generalized Aerodynamic Stiffness Matrix

	y	ϕ
y	0	$\bar{S}\left(-C_L - \alpha_1 C_{Y_\beta}\right)\alpha_1$
ϕ	0	$\bar{S}b\left(-C_{l_\beta}\cos\alpha_1 + C_{n_\beta}\sin\alpha_1\right)\alpha_1$
ψ	0	$\bar{S}b\left(-C_{n_\beta}\cos\alpha_1 - C_{l_\beta}\sin\alpha_1\right)\alpha_1$

	ψ	δ
y	$\bar{S}C_{Y_\beta}$	$-\bar{S}C_{Y_\delta}$
ϕ	$\bar{S}b\left(C_{l_\beta}\cos\alpha_1 - C_{n_\beta}\sin\alpha_1\right)$	$\bar{S}b\left(-C_{l_\delta}\cos\alpha_1 + C_{n_\delta}\sin\alpha_1\right)$
ψ	$\bar{S}b\left(C_{n_\beta}\cos\alpha_1 + C_{l_\beta}\sin\alpha_1\right)$	$\bar{S}b\left(-C_{n_\delta}\cos\alpha_1 - C_{l_\delta}\sin\alpha_1\right)$

Aerodynamic derivatives are determined for aircraft in either a flexibilized configuration, which includes the effects of static elastic doformation, or a truly rigid configuration which has no flexibility effects included. Since the actual aircraft is a flexible vehicle with often very significant flexibility effects on the dynamics, elastic effects should be incorporated to some degree. Aeroelastic increments to the rigid dynamics are usually calculated from a pre-determined elastic deformation and incorporated in the aerodynamic derivative formulation. In aeroelastic modeling, dynamics from the flexible mode degrees of freedom are coupled with the rigid body dynamics. Therefore, the static aeroelastic increments computed from this model are usually subtracted out if incorporating external flexibilized stability and control derivatives in the equations of motion [138].

Suppose the quasi-steady part of the state equation in terms of generalized coordinate η and control input δ,

Table 4.6: Lateral-directional Stability Axis Derivative Representation in the Generalized Aerodynamic Damping Matrix

	y	ϕ
y	$-\frac{\bar S}{U_1} C_{Y_\beta}$	$\frac{\bar S b}{2U_1}\left(-C_{Y_p} - \alpha_1 C_{Y_{\dot\beta}}\right)$
ϕ	$\frac{\bar S}{U_1} b\left(-C_{l_\beta}\cos\alpha_1\right)$ $+\frac{\bar S}{U_1} b\left(C_{n_\beta}\sin\alpha_1\right)$	$\frac{\bar S b}{2U_1} b\left(-C_{l_p}\cos\alpha_1\right)$ $+\frac{\bar S b}{2U_1} b\left(C_{n_p}\sin\alpha_1\right)$ $-\frac{\bar S b}{2U_1} b\left(C_{l_{\dot\beta}}(\cos\alpha_1)\alpha_1\right)$ $+\frac{\bar S b}{2U_1} b\left(C_{n_{\dot\beta}}(\sin\alpha_1)\alpha_1\right)$
ψ	$\frac{\bar S}{U_1} b\left(-C_{n_\beta}\cos\alpha_1\right)$ $-\frac{\bar S}{U_1} b\left(C_{l_\beta}\sin\alpha_1\right)$	$\frac{\bar S b}{2U_1} b\left(-C_{n_p}\cos\alpha_1\right)$ $-\frac{\bar S b}{2U_1} b\left(C_{l_p}\sin\alpha_1\right)$ $-\frac{\bar S b}{2U_1} b\left(C_{n_{\dot\beta}}(\cos\alpha_1)\alpha_1\right)$ $-\frac{\bar S b}{2U_1} b\left(C_{l_{\dot\beta}}(\sin\alpha_1)\alpha_1\right)$
	ψ	δ
y	$\frac{\bar S b}{2U_1}\left(-C_{Y_r} + C_{Y_{\dot\beta}}\right)$	0
ϕ	$\frac{\bar S b}{2U_1} b(-C_{l_r}\cos\alpha_1) + \frac{\bar S b}{2U_1} b(C_{n_r}\sin\alpha_1)$ $+\frac{\bar S b}{2U_1} b\left(C_{l_{\dot\beta}}\cos\alpha_1\right) - \frac{\bar S b}{2U_1} b\left(C_{n_{\dot\beta}}\sin\alpha_1\right)$	0
ψ	$\frac{\bar S b}{2U_1} b(-C_{n_r}\cos\alpha_1) - \frac{\bar S b}{2U_1} b(C_{l_r}\sin\alpha_1)$ $+\frac{\bar S b}{2U_1} b\left(C_{n_{\dot\beta}}\cos\alpha_1\right) + \frac{\bar S b}{2U_1} b\left(C_{l_{\dot\beta}}\sin\alpha_1\right)$	0

$$\eta = A\eta + B\delta, \tag{4.33}$$

is obtained as

$$\begin{aligned} A_{qs} &= A_{rr} + A_{ae} \\ B_{qs} &= B_{r\delta} + B_{ae} \end{aligned} \tag{4.34}$$

where $\{A_{rr}, B_{r\delta}\}$ are submatrices of $\{A, B\}$ corresponding to the rigid body degrees of freedom, r, and a set of control inputs, δ. Aeroelastic increments $\{A_{ae}, B_{ae}\}$ are obtained by static residualization

$$\begin{aligned} A_{ae} &= A_{re}(A_{ee})^{-1}A_{er} \\ B_{ae} &= A_{re}(A_{ee})^{-1}B_{e\delta} \end{aligned} \tag{4.35}$$

where $\{A_{re}, A_{ee}, A_{er}\}$ are submatrices of A associated with elastic dynamics, e, and rigid degrees of freedom, and $B_{d\delta}$ is defined similarly for controls. The nonlinear equations from a nonlinear simulation in terms of state variable, x, include the *flexibilized* aircraft nonlinear six degrees of freedom dynamics $\mathcal{A}_{nl}(x_{nl})$ with structural dynamic and control degrees of freedom

$$\dot x = \mathcal{A}_{nl} + \Delta\mathcal{A}_A x + \Delta\mathcal{A}_B\delta. \tag{4.36}$$

in which x combines x_{nl} with the elastic degrees of freedom, and δ are the input variables. To replace the quasi-steady aeroelastic portions of the $\{A, B\}$ matrices with the nonlinear equations, decompose the system matrices as

$$\begin{aligned} A &= A_{rr} + A_{dyn} \\ B &= B_{r\delta} + B_{dyn} \end{aligned} \tag{4.37}$$

with $\{A_{dyn}, B_{dyn}\}$ containing all rigid-elastic-control coupling degrees of freedom. From (4.34),

$$\begin{aligned} A &= A_{qs} - A_{ae} + A_{dyn} \\ B &= B_{qs} - B_{ae} + B_{dyn} \end{aligned} \tag{4.38}$$

and the static aeroelastic portions are subtracted from the coupled dynamics

$$\begin{aligned} \Delta\mathcal{A}_A &= A_{dyn} - A_{ae} \\ \Delta\mathcal{A}_B &= B_{dyn} - B_{ae}. \end{aligned} \tag{4.39}$$

to give the linear dynamic increments $\{\Delta\mathcal{A}_A, \Delta\mathcal{A}_B\}$ of the nonlinear simulation equations (4.36). Alternatively, if the aerodynamic derivatives do not include flexibility effects, the static aeroelastic increments from the model are retained.

4.3 Servoelastic Modeling

Aeroservoelastic analysis requires accurate description of dynamics over a frequency range dependent on the dynamics of the controller. Dynamics include the coupling of structural dynamics, control dynamics, feedback sensors, aerodynamics, and actuation.

Extensive effort is usually employed in structural modeling and validation with ground test data to minimize errors in modal dynamics. Mode shape errors still tend to persist as a major source of uncertainty for sensor feedbacks. Other sources of errors in control feedback sensing are fairly well understood. Sensor calibration minimizes bias errors, and noise properties are generally specified in sensor noise-to-signal ratios. Aerodynamic modeling in the aeroelastic equations of motion is a broad and sensitive topic, depending on type of flow regime (subsonic, transonic, or supersonic) and solution method. Computational fluid dynamics will remain an active field of research in this area and will not be discussed. Suffice it to mention that errors will exist in unsteady aerodynamic solutions and also in coupling aerodynamics with elastic and control dynamics [22, 111], especially in sampled-data and multirate feedback control systems.

Control dynamics are well understood as a separate entity since they are designed, tested, and extensively verified before flight. Uncertainty in feedback is less deterministic. Closed loop effects of uncertainty on stability and performance from noise, parametric uncertainty, nonlinearities, sampled-data, multirate sampling, and other modeling errors can be quite complicated. Linear analysis tools [5] are available to estimate these effects, yet to some extent they depend on model development with the proper coupling mechanisms between aerodynamics, elasticity, and control. These must be accurately modeled over a designated frequency range. Sampled-data systems become particularly troublesome to model because of high frequency aeroelastic dynamics and their characteristics due to the Nyquist effect, such as aliasing. Discretization needs to be done carefully since the analysis may be significantly affected by coupling between aeroelastic and control system dynamics.

Actuator responses are customarily represented with transfer functions matched to test results, and control surface stiffness is commonly modeled as a linear spring in the formulation of the airframe dynamics [145]. For high bandwidth flight control systems it is preferable to improve the actuation modeling by considering the physical properties of the system and its compliance with the structure. More realistic predictions of airframe-actuation coupling, consistent with advances in modeling the airframe and control system, can be achieved by considering the physical properties of the system and its installation on the airframe. Detailed models based on electrical, hydraulic, and mechanical characteristics, with complex impedance effects, are desirable for more fidelity.

Reference [37] emphasizes, by example, that detailed models based on electrical and hydromechanical elements are critical for accuracy. The concern is highlighted in [21] where the complexity of actuation coupling mechanisms significantly affects a fighter aircraft aeroservoelastic stability prediction.

Methods in [125, 126] include detailed nonlinear actuation system models in aeroservoelastic simulation analysis. For a linear analysis of the actuation system the following assumptions are

- linear valve flow
- no valve overlaps or underlaps
- parallel-sided valve port
- equal fluid volumes on each side of main ram pistons
- equal areas on each side of main ram pistons
- no steady load on ram
- no friction, backlash, or hysteresis
- no limits encountered.

Small amplitude characteristics are affected by overlap, hysteresis, friction, and backlash. Limits, port shape, and ram symmetry affect large amplitude motion. Therefore the model is presumed to be most accurate for medium range amplitudes.

Modeling begins by separating the elements of the actuation system into a plant, which contains mechanical and hydraulic components, and a feedback system which contains servovalve and main ram feedbacks. The perturbation of a free control surface m_o couples with the modal dynamics according to the stiffness k_1, damping b_1, and modal mass m_1 of a particular structural mode through the hinge moment. Aerodynamic forces with components of damping and stiffness k_0, furthermore, induce aerodynamic hinge moments. The actuator is attached to the control surface by a lever arm and to the airframe with some mounting assembly. The motion of the piston relative to the main ram body, x_p, is fed back to the commanded signal x_i. Feedback is also employed in the servovalve.

The servovalve dynamics are combined with an inner loop position feedback. The outer loop ram position feedback controls the ram position according to external demand x_i from the control system. Dynamic actuator stiffness is a sum of the airframe attachment stiffness and the control surface lever arm attachment stiffness. Damping comes from these sources as well as structural mode, aerodynamic, and hydraulic origins. External force F_e is designated as acting on the main mass from a ground impedance test, for instance, to check stiffness.

Derivation of linear actuation models is valid for multiple hydraulic systems operating the same actuator in tandem or parallel. The systems are assumed identical and the output force produced by the ram is multiplied by n (usually 1 or 2), the number of operating systems. For this analysis, b_t is the total damping contribution from aerodynamic, structural (b_1), and main ram damping. Masses of the main ram body and piston are neglected since they are small compared to control surface inertia, so the dampings in the actuator attachment and output lever arm structures are also neglected. In matrix form the main ram hydromechanical dynamic equations then become

$$\begin{pmatrix} \dot{P_j} \\ \dot{x_0} \\ \ddot{x_0} \\ \dot{x_1} \\ \ddot{x_1} \end{pmatrix} = \begin{bmatrix} -C_p & 0 & -\frac{A_p}{C_c} & 0 & 0 \\ 0 & 0 & 1 & 0 & 0 \\ \frac{nA_p}{m_0} & -\frac{k_0+k_1}{m_0} & -\frac{b_t}{m_0} & \frac{k_1}{m_0} & \frac{b_1}{m_0} \\ 0 & 0 & 0 & 0 & 1 \\ 0 & \frac{k_1}{m_1} & \frac{b_1}{m_1} & -\frac{k_1}{m_1} & -\frac{b_1}{m_1} \end{bmatrix} \begin{pmatrix} P_j \\ x_0 \\ \dot{x_0} \\ x_1 \\ \dot{x_1} \end{pmatrix} + \begin{bmatrix} \frac{C_v}{C_c} & 0 \\ 0 & 0 \\ 0 & \frac{1}{m_0} \\ 0 & 0 \\ 0 & 0 \end{bmatrix} \begin{pmatrix} x_v \\ F_e \end{pmatrix} \tag{4.40}$$

$$\begin{pmatrix} x_p \\ x_0 \end{pmatrix} = \begin{bmatrix} \frac{nA_p}{k_T} & 1\,0\,0\,0 \\ 0 & 1\,0\,0\,0 \end{bmatrix} \begin{pmatrix} P_j \\ x_0 \\ \dot{x_0} \\ x_1 \\ \dot{x_1} \end{pmatrix} + \begin{bmatrix} 0\,0 \\ 0\,0 \end{bmatrix} \begin{pmatrix} x_v \\ F_e \end{pmatrix} \tag{4.41}$$

for a sign convention assuming that actuator *extension* always gives *negative* hinge moment and control rotation. Therefore, for a servovalve command x_v or applied test force F_e, the ram piston is displaced x_p relative to the ram body. These equations include the effects of coupling with a structural mode of generalized mass m_1, and aerodynamic coupling of the surface inertia through the m_0 terms. The constant C_p is a composite coefficient which may be interpreted as a combination of servovalve flow pressure gain and some leakage across the piston head.

This main ram model is augmented with the servovalve feedback and feedforward (filter time constants τ_1 and τ_2, respectively) and ram feedback (time constant τ_3). For s_i states corresponding to respective time constants τ_i, and X denoting a nonzero value which is a function of the system parameters, the equations become (δ_d is actuator command)

$$\begin{pmatrix} \dot{s}_2 \\ \dot{s}_1 \\ \dot{s}_0 \\ \dot{P}_j \\ \dot{x}_0 \\ \ddot{x}_0 \\ \dot{x}_1 \\ \ddot{x}_1 \\ \dot{s}_3 \end{pmatrix} = \begin{bmatrix} -\frac{1}{\tau_2} & 0 & 0 & 0 & 0 & 0 & 0 & 0 & X \\ X & -\frac{1}{\tau_1} & X & 0 & 0 & 0 & 0 & 0 & 0 \\ 0 & 1 & 0 & 0 & 0 & 0 & 0 & 0 & 0 \\ 0 & 0 & X & -C_p & 0 & -\frac{A_p}{C_c} & 0 & 0 & 0 \\ 0 & 0 & 0 & 0 & 0 & 1 & 0 & 0 & 0 \\ 0 & 0 & 0 & \frac{nA_p}{m_0} & -\frac{k_0+k_1}{m_0} & -\frac{b_t}{m_0} & \frac{k_1}{m_0} & \frac{b_1}{m_0} & 0 \\ 0 & 0 & 0 & 0 & 0 & 0 & 0 & 1 & 0 \\ 0 & 0 & 0 & 0 & \frac{k_1}{m_1} & \frac{b_1}{m_1} & -\frac{k_1}{m_1} & -\frac{b_1}{m_1} & 0 \\ 0 & 0 & 0 & X & X & 0 & 0 & 0 & -\frac{1}{\tau_3} \end{bmatrix} \begin{pmatrix} s_2 \\ s_1 \\ s_0 \\ P_j \\ x_0 \\ \dot{x}_0 \\ x_1 \\ \dot{x}_1 \\ s_3 \end{pmatrix} - x_h \delta_d \tag{4.42}$$

$$\begin{pmatrix} \delta \\ \dot{\delta} \\ x_p \end{pmatrix} = \begin{bmatrix} 0 & 0 & 0 & 0 & -\frac{1}{x_h} & 0 & 0 & 0 & 0 \\ 0 & 0 & 0 & 0 & 0 & -\frac{1}{x_h} & 0 & 0 & 0 \\ 0 & 0 & 0 & \frac{nA_p}{k_T} & 1 & 0 & 0 & 0 & \end{bmatrix} \begin{pmatrix} s_2 \\ s_1 \\ s_0 \\ P_j \\ x_0 \\ \dot{x}_0 \\ x_1 \\ \dot{x}_1 \\ s_3 \end{pmatrix} + 0 \cdot \delta_d. \tag{4.43}$$

Given the extensive assumptions for linearity and ignoring other system components such as a dynamic leak valve, damper accumulator, or some other pressure feedback modulation, the model is not expected to match test results exactly. A model for friction would generally add some damping. Also, the fluid coefficients C_p, C_v, and C_c are susceptible to error.

These are parameters that are uncertain and can be related to uncertain matrix elements for actuator model validation. For example, the main valve flow coefficient C_v indicates the flow through the ports to the load per unit displacement of the valve spool. It assumes parallel-sided ports, zero laps, and no leakage. It determines the flow gain between valve movement and the motion response of the load piston, which is inversely proportional to the main ram piston area. Fluid flow pressure constant C_p, inversely proportional to C_c, is the cross piston leak flow coefficient. These often need to be adjusted to account for variations in valve port shape, size, and pressure difference for linearization of cross piston leak flow. The fluid bulk modulus may differ depending on the type of operating system [37], which affects C_c and consequently C_p.

Parametric uncertainty operators (discussed in the context of uncertainty descriptions in a later chapter) are augmented with these terms to account for unknown bounded variations. Other uncertainties due to nonlinearity (friction, hysteresis) and neglected terms in the model can also be accounted for with dynamic uncertainty operators. Uncertainty operators are weighted and a model validation test performed with actuator test data to produce a robust actuator model to incorporate in a robust aeroservoelastic model.

Discounting the states due to structural and control surface inertias, these high order actuator models can easily get to be near 10th order, depending predominantly on the type of servovalve and the inner loop servovalve feedback mechanism. Alternatively, low order models conventionally used for aeroservoelastic analysis are typically 2nd to 4th order, with a corresponding frequency used to define the dynamic stiffness between the airframe and control surface. They are generated by simply fitting a low order equivalent transfer function to measured data, with no regard for actuator details. Compliance with the structure is provided by the stiffness frequency, defined as $\omega = \sqrt{k_T/m_o}$, which is derived from a measured dynamic stiffness k_T and surface mass m_0. However, low order models can give extremely unreliable stability and performance estimates [21].

The structural model includes rigid body, flexible, and control surface modal dynamics with cross inertia terms. In essence, disregarding aerodynamic lag states (for simplicity in demonstration), the state equations take the following form

$$\begin{pmatrix} \dot{\eta}_r \\ \dot{\eta}_e \\ \ddot{\eta}_r \\ \ddot{\eta}_e \end{pmatrix} = [A] \begin{pmatrix} \eta_r \\ \eta_e \\ \dot{\eta}_r \\ \dot{\eta}_e \end{pmatrix} + [B\ \acute{B}] \begin{pmatrix} \delta \\ \hline \dot{\delta} \end{pmatrix} \tag{4.44}$$

$$y = [C] \begin{pmatrix} \eta_r \\ \eta_e \\ \dot{\eta}_r \\ \dot{\eta}_e \end{pmatrix} + [D\ \acute{D}] \begin{pmatrix} \delta \\ \hline \dot{\delta} \end{pmatrix} \tag{4.45}$$

for rigid body states η_r, flexible mode states η_e, control surface displacements δ, and control surface rates $\dot{\delta}$. Actuator states are augmented by adjusting this matrix quadruple as

$$\begin{pmatrix} \dot{\eta}_r \\ \dot{\eta}_e \\ \ddot{\eta}_r \\ \ddot{\eta}_e \\ \hline \dot{\delta} \end{pmatrix} = \begin{bmatrix} A & B \\ 0 & 0 \end{bmatrix} \begin{pmatrix} \eta_r \\ \eta_e \\ \dot{\eta}_r \\ \dot{\eta}_e \\ \hline \delta \end{pmatrix} + \begin{bmatrix} \acute{B} \\ I \end{bmatrix} (\dot{\delta}) \tag{4.46}$$

$$y = [C\ D] \begin{pmatrix} \eta_r \\ \eta_e \\ \dot{\eta}_r \\ \dot{\eta}_e \\ \hline \delta \end{pmatrix} + [\acute{D}]\,(\dot{\delta})\,. \tag{4.47}$$

The equation for control surface rate $\dot{\delta}$ is augmented to this model for each actuator to produce

$$\begin{aligned} \dot{x} &= \bar{A}x + \bar{B}\delta_d \\ y &= \bar{C}x + \bar{D}\delta_d \end{aligned} \tag{4.48}$$

where

$$x = \lceil \eta_a \eta_r \eta_e \dot{\eta}_r \dot{\eta}_e \delta \rfloor \tag{4.49}$$

includes all actuator states η_a. This continuous-time aircraft plant description is augmented with either analog or more prevalent discrete control system dynamics to construct the aeroservoelastic model.

Aeroservoelastic model realization with the control system differs according to the desired goals. References [134, 135] model sampled-data (continuous-discrete) systems subject to arbitrary delays and multiple (possibly asynchronous) sample rates for frequency response analyses. Other formulations of sampled-data and multirate systems can be found in [29, 43, 140]. Time response and eigenvalue analyses are performed by discretizing the plant at the highest loop sample rate $\frac{1}{T}$, using the method of zero-order hold (ZOH) equivalence (matrix exponential), and incorporating control processor time delay $\tau = mT - \lambda$, where T is the sample time, m is an integer, and λ a fractional delay difference (assuming $m \geq 0$ and $0 \leq \lambda < T$). Treatment of the case $m = 0, \tau = -\lambda$ is actually *prediction* and will not be discussed. Here we consider only the case $m = 1, \lambda \neq 0$. First, define

$$\Phi = e^{\bar{A}T},\ \Gamma_1 = \textstyle\int_\lambda^T e^{\bar{A}t}\bar{B}\ dt,\ \Gamma_2 = \displaystyle\int_0^\lambda e^{\bar{A}t}\bar{B}\ dt. \tag{4.50}$$

For any positive scalar a define

$$\Phi(a) = e^{\bar{A}a}\ ,\ \Psi(a) = \frac{1}{a}\int_0^a e^{\bar{A}t}dt. \tag{4.51}$$

Next write

$$\Gamma_1 = \Phi(\lambda)\Psi(T-\lambda)(T-\lambda)\bar{B} \tag{4.52}$$

$$\Gamma_2 = \Psi(\lambda)\lambda\bar{B}\ . \tag{4.53}$$

Only the exponential series for Ψ need be computed since it can be shown that

$$\Phi(a) = I + \Psi(a)a\bar{A} \ . \tag{4.54}$$

The state equations are now given by

$$x(k+1) = \Phi x(k) + \Gamma_1 u(k-1) + \Gamma_2 u(k). \tag{4.55}$$

Define the new state $x_{n+1}(k) = u(k-1)$ to form the new system

$$\begin{pmatrix} x(k+1) \\ x_{n+1}(k+1) \end{pmatrix} = \begin{bmatrix} \Phi & \Gamma_1 \\ 0 & 0 \end{bmatrix} \begin{pmatrix} x(k) \\ x_{n+1}(k) \end{pmatrix} + \begin{bmatrix} \Gamma_2 \\ I \end{bmatrix} u(k) \tag{4.56}$$

$$y = [\bar{C} \ 0] \begin{pmatrix} x \\ x_{n+1} \end{pmatrix} + \bar{D}u(k). \tag{4.57}$$

The formulation therefore adds delayed control inputs to the state vector for delay times less than the smallest loop sample time,. It is computationally attractive and no approximation is involved as in the common technique of using low-order Padé approximates for delays. This is the plant description for computation of time responses and open loop eigenvalues.

Zero-order holds and time delays are treated differently in frequency response analyses. Plant dynamics are Laplace-transformed and augmented with analog elements, including sensors and prefilters, using standard procedures. Then the continuous system with time delay τ is formulated as

$$\begin{aligned} y(s) &= \left(\frac{H(s)[ZOH]}{T} \right) u(s) \\ &= H^*(s, \tau, T)u(s) \end{aligned} \tag{4.58}$$

where s is the Laplace variable, and

$$H(s) = \left[\bar{C} \ (sI - \bar{A}\,)^{-1} \bar{B} \ + \bar{D} \ \right] \tag{4.59}$$

$$[ZOH] = e^{-s\tau} \frac{1 - e^{-sT}}{s} \tag{4.60}$$

where $H^*(s, \tau, T)$ is now the plant transfer function for all continuous dynamics.

Feedback control laws often consist of a mixture of sampling rates, so a procedure for modeling sampled-data multirate systems for frequency response analyses is required [29, 43]. In the sampled-data model, assume the feedbacks are sampled at two different rates $\frac{1}{T_3}$ and $\frac{1}{T_4}$, the reference and controlled variable error signals sampled at $\frac{1}{T}$, and control surfaces commanded at possibly two different rates $\frac{1}{T_1}$ and $\frac{1}{T_2}$ designated in the zero-order holds. The $\frac{1}{T}$ factor in the zero-order holds is derived from taking the limit of a sampled spectrum as the sample time goes to zero, thereby resulting in the continuous spectrum of the discretely excited plant. Aircraft dynamics are described by Laplace transformed state equations, so system transfer functions are all well-defined functions of the Laplace variable s similar to the single-rate sampled-data formulation.

Eigenvalue and time response analyses require plant and controller to be discretized with an identical sample time. Hence, feedbacks sampled at relatively slower rates than other faster rate feedbacks are converted to faster signals using discrete memory averagers. For the case of a faster-rate to slower-rate ratio of n=2:1, the averager is $M(\bar{z}) = \frac{1+\bar{z}}{2\bar{z}}$, where $\bar{z} = e^{sT}$. The multirate character of feedback is disguised and all digital filters can be implemented at the higher rate discretization. Averagers are generally represented by

$$M(\bar{z}) = \frac{\sum_{k=0}^{n} \bar{z}^k}{(n+1)\bar{z}^n} \tag{4.61}$$

where the rate ratio $n = \frac{T_i}{T}, T_i > T$ is an integer. In the discrete multirate implementation for eigenvalue analyses, plant dynamics are discretized at the fastest rate $\frac{1}{T}$ using zero-order hold equivalence, and slower-rate feedbacks are passed through the appropriate $M(\bar{z})$. Any filters calculated and implemented for slower rate signals substitute $\bar{z} = e^{sT_i}$. Slow rate control commands are also averaged before the actuators.

4.4 Aeroelastic and Aeroservoelastic Stability

Aeroservoelasticity is inherently the discipline of aeroelasticity including the control system dynamics (sensors, control laws, and actuators), and the classical aeroelastic equations of motion are certainly an integral part of the aeroservoelastic model. Aeroelasticity is in itself a feedback problem between structural dynamics and unsteady aerodynamics. It is a conservative elastomechanical system. However, aeroservoelastic dynamics can have properties very distinct from the field of aeroelasticity because of control surface motion, actuation, modal contributions to sensor dynamics, and active feedback control dynamics which tend to "distort" the natural aeroelastic properties with sufficient control bandwidth. Feedback introduces energy to the aircraft through the actuators, which are often controlled by internal feedback mechanisms.

The field of aeroservoelasticity emerged from the intersection of essentially independent aircraft design considerations. Increases in aircraft strength-to-weight ratio with more structural flexibility conflicted with increasing bandwidth and higher authority control elements in the actuators and control dynamics. Eventually the interactions coupled significantly to create adversity [40]. In modern systems the distinction between traditional flight control and aeroelastic control is becoming less obvious [83, 100, 139]. Whether the objective is to actively control the elastic shape or structural dynamics, or passively remove the structural interactions with filtering, an aeroservoelastic system is hereby defined as an *aeroelastic system with feedback control.*

This section will be a brief treatment of flutter analyses, computational approaches in aeroelasticity, extensions and modifications for aeroservoelastic analysis, and aeroservoelastic stability prediction. First write the flutter equations in a mixed (function of time and harmonic frequency) matrix form as

$$M\ddot{x} + Kx = \bar{q}[A_e(M_\infty, k)]x \tag{4.62}$$

for modal displacements x, mass M and stiffness K, aerodynamic influence coefficients $[A_e(M_\infty, k)]$ as functions of freestream Mach number M_∞ and reduced frequency k, and dynamic pressure $\bar{q}$. Unsteady forces A_e assume sinusoidal motion in terms of reduced frequency. Transformation to generalized coordinates η from physical coordinates x through the mode shape matrix T, by $x = T\eta$, reduces the problem considerably to the distributed natural modal space and allows for computational efficiency. Then assuming harmonic motion $x = |x|e^{\jmath\omega t}$ gives

$$M\ddot{\eta} + K\eta = \bar{q}[A_e(M_\infty, k)]\eta. \tag{4.63}$$

Aerodynamic damping is represented in the influence coefficients, but structural damping force is usually imposed by multiplying the structural stiffness by the factor $\jmath g$, thereby making the force 90 degrees out-of-phase with displacement. Structural damping coefficient g is generally assumed to be about 0.02 for all modes, and a factor $(1 + \jmath g)$ is applied to the stiffness matrix written as $K = [\omega_j^2 M_{jj}]$ in terms of the modal natural frequencies ω_j. The flutter equation is now

$$M\ddot{\eta} + (1 + \jmath g)[\omega_j^2 M_{jj}]\eta = \bar{q}[A_e(M_\infty, k)]\eta. \tag{4.64}$$

Algorithms for solving the flutter equation can be categorized according to the assumed functional dependence of the aerodynamics and the assumed motion characteristics [34]. Three primary procedures assume the motion is of the form $\eta = |\eta|e^{pt}, p = \gamma + \jmath\omega$. The *V-g* method assumes sinusoidal motion of both aerodynamics and structure, or $\eta = |\eta|e^{\jmath\omega t}$. The advantage is that the aerodynamic forces are only determined for real frequencies, but there is lack of physical insight in the interpretation of the results. The *p*-method assumes damped motion $\eta = |\eta|e^{pt}$, $p = \gamma + \jmath\omega$, and the aerodynamics are computed using a complex reduced frequency. Finally, the *p-k* method also assumes damped motion $\eta = |\eta|e^{pt}$ but the aerodynamics are computed with real reduced frequency.

From the complex flutter equation (4.64), the *V-g* method solves for frequency, so the equation is modified as

$$\{-\omega^2 M + (1 + \jmath g)[\omega_j^2 M_{jj}] - \bar{q}[A_e(M_\infty, k)]\}\eta = 0$$

$$\{M - (1 + \jmath g)\frac{1}{\omega^2}[\omega_j^2 M_{jj}] + \bar{q}\frac{\bar{b}^2}{\bar{b}^2\omega^2}[A_e(M_\infty, k)]\}\eta = 0 \tag{4.65}$$

where $\bar{b}^2$ ($\bar{b} = \frac{\bar{c}}{2}$) is introduced to form reduced frequency. Divide through by a reference mass, m_o, and combine with density to form a mass ratio. Let $m_o\bar{M} = M$, define the mass ratio

$$\varrho = \frac{m_o}{\pi \rho \bar{b}^2} \tag{4.66}$$

to get

$$\{M - (1 + \jmath g)\frac{1}{\omega^2}[\omega_j^2 M_{jj}] + \frac{1}{\pi \varrho k^2}[A_e(M_\infty, k)]\}\eta = 0. \tag{4.67}$$

The eigenvalue

$$\chi = \chi_r + \jmath\chi_i = (1 + \jmath g)\frac{1}{\omega^2} \tag{4.68}$$

has the associated eigenfrequency, $\omega = \frac{1}{\sqrt{\chi_r}}$, velocity,

$$V = \frac{\omega \bar{b}}{k} = \frac{\bar{b}}{k\sqrt{\chi_r}}, \tag{4.69}$$

and damping $g = \frac{\chi_i}{\chi_r}$. This value of damping is that which must be added to the system to get pure sinusoidal motion. At fixed M_∞ and ρ, k is varied and damping is tracked versus velocity until $g \leq 0$.

Predicted flutter conditions in which V, M_∞, and ρ for a given pressure altitude are in simultaneous agreement with the standard atmosphere are called *matched* flutter points. Standard flutter analysis algorithms use M_∞ and ρ as inputs, but the flutter velocity is not matched to the standard atmosphere from the specified M_∞ and ρ. This necessitates a flutter prediction algorithm that incorporates additional iterations to find flutter solutions along a standard atmosphere curve.

Non-zero damping solutions from the three procedures will be different, which may be very significant. In the case of so-called "hump" modes, a mode crosses into an unstable region, as a function of equivalent velocity (velocity at sea level with $\bar{q}$ matching velocity at altitude), and returns into the stable region at a higher velocity. Usually this indicates a mild onset into flutter, or at best, disturbing ride qualities. In general the behavior of a flutter solution just before instability may be an important property to monitor.

Several types of flutter are possible. Most common is the bending-torsion variety, or frequency-coalescence flutter, in which two or more modal frequencies, χ_r, come together near instability, and one modal damping χ_i increases rapidly while the other decreases to instability. Out-of-phase damping forces are relatively unimportant to the coupling instability between pitch (torsion) and plunge (bending). In another type, a single degree of freedom phenomena is realized where modal frequency χ_r remains constant, and damping χ_i slowly decreases to instability. This type is very sensitive to damping forces. Finally, a static instability can be realized at zero frequency, called *divergence*.

Inclusion of actuation and flight controls for aeroservoelasticity [145] imposes modified modeling formulations into a state-space framework for control applications. The traditional aeroelastic dependence on reduced frequency is transformed to explicit rational functions of the states. Aerodynamic forces are fit with rational function approximations in the frequency domain, and these approximations are then transformed to the time domain using the Laplace transform. Excellent descriptions of the methods are contained in [128, 129]. Efficiency and accuracy have been improved upon [64], and time series approaches [101] have been successfully attempted. The more recent method [138] avoids the extra aerodynamic lag states in the state space approximation by using a *p-k* flutter solution process to estimate linear approximations to the aerodynamic forces at flutter eigenvalue frequencies.

Aeroservoelastic instability mechanisms are more obtuse than traditional flutter. Control system loop gains change constantly with flight condition, as do aerodynamic hinge moments and effective actuator bandwidths. Variations in the interaction between aeroelastic and control system dynamics from aircraft weight variations and distribution, aerodynamics from different flight regimes, and control allocation changes are difficult to encompass with standard analyses. Propulsive forces may be significant with thrust-vectored configurations. Aerothermodynamics change aeroelastic properties in hypersonic flight regimes. Aeroservoelastic stability analysis can produce unexpected results from the increased complexity of the coupling dynamics, especially when uncertainty is considered [25, 23].

As computational fluid dynamics (CFD) codes progress in speed and accuracy, aeroelastic models can be derived from CFD simulations [55]. Simulation of aeroelastic dynamics are used to produce time responses for system identification techniques often used for data analysis. Identified state space solutions are then extrapolated for flutter prediction [31], or models are constructed directly from CFD time responses for control applications [112, 113]. Control design and aeroservoelastic analysis are quickly becoming feasible in this framework. Complete aeroservoelastic dynamical simulations with actuator dynamics, multirate sampled-data control systems, and sensor dynamics, complete with modeled nonlinearities, are quite practical once the aeroelastic dynamics can be solved with sufficient accuracy and efficiency.

A.4 Inertial to Body-fixed Coordinate Transformation

The reference axes used to compute the aerodynamics resulting from the free vibration modes of unconstrained elastic bodies is the inertial frame. However for stability and control analysis the prefered coordinate system is the body fixed coordinate system. If the inertial frame and reference axes are inclined with respect to each other *in a trim state* through the Euler angles (Ψ, Θ, Φ), then the velocities at the aircraft center-of-gravity in the inertial frame and body-fixed reference axes are related by

$$\begin{pmatrix} \dot{X} \\ \dot{Y} \\ \dot{Z} \end{pmatrix}_I = \Gamma(\Psi, \Theta, \Phi) \cdot \begin{pmatrix} U \\ V \\ W \end{pmatrix}_B \tag{A.4.1}$$

in which

$$\Gamma = \begin{bmatrix} \cos\Psi\cos\Theta & \cos\Psi\sin\Theta\sin\Phi - \sin\Psi\cos\Phi & \cos\Psi\sin\Theta\cos\Phi + \sin\Psi\sin\Phi \\ \sin\Psi\cos\Theta & \sin\Psi\sin\Theta\sin\Phi + \cos\Psi\cos\Phi & \sin\Psi\sin\Theta\cos\Phi - \cos\Psi\sin\Phi \\ -\sin\Theta & \cos\Theta\sin\Phi & \cos\Theta\cos\Phi \end{bmatrix} \tag{A.4.2}$$

Similarly, the accelerations in the inertial frame are given by

$$\begin{aligned} \begin{pmatrix} \ddot{X} \\ \ddot{Y} \\ \ddot{Z} \end{pmatrix}_I &= \Gamma(\Psi, \Theta, \Phi) \cdot \left[\begin{pmatrix} \dot{U} \\ \dot{V} \\ \dot{W} \end{pmatrix} + \Omega \cdot \begin{pmatrix} U \\ V \\ W \end{pmatrix} \right]_B \\ &= \Gamma(\Psi, \Theta, \Phi) \cdot \begin{pmatrix} A_x \\ A_y \\ A_z \end{pmatrix}_B \end{aligned} \tag{A.4.3}$$

where Ω is the skew-symmetric angular velocity matrix. Considering the accelerations in the body coordinate system

$$\begin{aligned} \begin{pmatrix} A_x \\ A_y \\ A_z \end{pmatrix}_B &= \begin{pmatrix} \dot{U} \\ \dot{V} \\ \dot{W} \end{pmatrix} + \Omega \cdot \begin{pmatrix} U \\ V \\ W \end{pmatrix} \\ &= \begin{pmatrix} \dot{U} \\ \dot{V} \\ \dot{W} \end{pmatrix} + \begin{bmatrix} 0 & -R & Q \\ R & 0 & -P \\ -Q & P & 0 \end{bmatrix} \begin{pmatrix} U \\ V \\ W \end{pmatrix} \\ &= \begin{pmatrix} \dot{U} \\ \dot{V} \\ \dot{W} \end{pmatrix} + \begin{bmatrix} -RV + QW \\ RU - PW \\ -QU + PV \end{bmatrix} . \end{aligned} \tag{A.4.4}$$

Now with reference to a trim state at orientation $(\Psi_1, \Theta_1, \Phi_1)$,

$$
\begin{aligned}
\Phi &= \Phi_1 \\
\Theta &= \Theta_1 \\
\Psi &= \Psi_1 \\
U &= U_1 \\
V &= V_1 \\
W &= W_1 \\
P &= P_1 \\
Q &= Q_1 \\
R &= R_1
\end{aligned}
\tag{A.4.5}
$$

the perturbations are

$$
\begin{aligned}
\Phi &= \Phi_1 + \phi \\
\Theta &= \Theta_1 + \theta \\
\Psi &= \Psi_1 + \psi \\
U &= U_1 + u \\
V &= V_1 + v \\
W &= W_1 + w \\
P &= P_1 + p \\
Q &= Q_1 + q \\
R &= R_1 + r \quad .
\end{aligned}
\tag{A.4.6}
$$

The perturbed velocities in the inertial frame become

$$
\begin{pmatrix} \dot{x} \\ \dot{y} \\ \dot{z} \end{pmatrix} = \Gamma_1 \cdot \begin{pmatrix} u \\ v \\ w \end{pmatrix} + \tilde{\Lambda}_1 \cdot \begin{pmatrix} \phi \\ \theta \\ \psi \end{pmatrix}
\tag{A.4.7}
$$

for $\Gamma_1 = \Gamma(\Psi_1, \Theta_1, \Psi_1)$, and

$$
\tilde{\Lambda}_1 = \begin{bmatrix} \Lambda_{x,\phi} & \Lambda_{x,\theta} & \Lambda_{x,\psi} \\ \Lambda_{y,\phi} & \Lambda_{y,\theta} & \Lambda_{y,\psi} \\ \Lambda_{z,\phi} & \Lambda_{z,\theta} & \Lambda_{z,\psi} \end{bmatrix} .
\tag{A.4.8}
$$

The elements of $\tilde{\Lambda}_1$ are as follows.

$$
\begin{aligned}
\Lambda_{x,\phi} &= V_1(\cos\Phi_1 \sin\Theta_1 \cos\Psi_1 + \sin\Phi_1 \sin\Psi_1) \\
&\quad + W_1(\cos\Phi_1 \sin\Psi_1 - \sin\Phi_1 \sin\Theta_1 \cos\Psi_1) \\
\Lambda_{x,\theta} &= -U_1 \sin\Theta_1 \cos\Psi_1 + V_1 \sin\Phi_1 \cos\Theta_1 \cos\Psi_1 \\
&\quad + W_1 \cos\Phi_1 \cos\Theta_1 \cos\Psi_1 \\
\Lambda_{x,\psi} &= -U_1 \cos\Theta_1 \sin\Psi_1 - V_1(\sin\Phi_1 \sin\Theta_1 \sin\Psi_1 + \cos\Phi_1 \cos\Psi_1)
\end{aligned}
$$

$$
\begin{aligned}
&\quad + W_1(\sin\Phi_1 \cos\Psi_1 - \cos\Phi_1 \sin\Theta_1 \sin\Psi_1)\\
\Lambda_{y,\phi} &= V_1(\cos\Phi_1 \sin\Theta_1 \sin\Psi_1 - \sin\Phi_1 \cos\Psi_1\\
&\quad - W_1(\sin\Phi_1 \sin\Theta_1 \sin\Psi_1 + \cos\Phi_1 \cos\Psi_1)\\
\Lambda_{y,\theta} &= -U_1 \sin\Theta_1 \sin\Psi_1 + V_1 \sin\Phi_1 \cos\Theta_1 \sin\Psi_1\\
&\quad + W_1 \cos\Phi_1 \cos\Theta_1 \sin\Psi_1\\
\Lambda_{y,\psi} &= U_1 \cos\Theta_1 \cos\Psi_1\\
&\quad + V_1(\sin\Phi_1 \sin\Theta_1 \cos\Psi_1 - \cos\Phi_1 \sin\Psi_1)\\
&\quad + W_1(\cos\Phi_1 \sin\Theta_1 \cos\Psi_1 + \sin\Phi_1 \sin\Psi_1)\\
\Lambda_{z,\phi} &= V_1 \cos\Phi_1 \cos\Theta_1 - W_1 \sin\Phi_1 \cos\Theta_1\\
\Lambda_{z,\theta} &= -U_1 \cos\Theta_1 - V_1 \sin\Phi_1 \sin\Theta_1 - W_1 \cos\Phi_1 \sin\Theta_1\\
\Lambda_{z,\psi} &= 0.
\end{aligned}
\tag{A.4.9}
$$

These equations are valid for general reference conditions with the inertia axes initially oriented to coincide with the body axes through Γ_1.

The body axis perturbation accelerations are

$$
\begin{pmatrix} a_x \\ a_y \\ a_z \end{pmatrix}_B = \begin{pmatrix} \dot{u} \\ \dot{v} \\ \dot{w} \end{pmatrix} + \tilde{\Omega} \cdot \begin{pmatrix} u \\ v \\ w \end{pmatrix} + \tilde{\Lambda} \cdot \begin{pmatrix} p \\ q \\ r \end{pmatrix} \tag{A.4.10}
$$

where

$$
\tilde{\Omega} = \begin{bmatrix} 0 & -R_1 & Q_1 \\ R_1 & 0 & -P_1 \\ -Q_1 & P_1 & 0 \end{bmatrix} \tag{A.4.11}
$$

and

$$
\tilde{\Lambda} = \begin{bmatrix} 0 & W_1 & -V_1 \\ -W_1 & 0 & U_1 \\ V_1 & -U_1 & 0 \end{bmatrix}. \tag{A.4.12}
$$

Perturbed accelerations in the inertial frame then become

$$
\begin{aligned}
\begin{pmatrix} \ddot{x} \\ \ddot{y} \\ \ddot{z} \end{pmatrix}_I &= \Gamma_1 \cdot \begin{pmatrix} a_x \\ a_y \\ a_z \end{pmatrix}_B \\
&= \Gamma_1 \cdot \begin{pmatrix} \dot{u} \\ \dot{v} \\ \dot{w} \end{pmatrix} + \Gamma_1 \cdot \tilde{\Omega} \cdot \begin{pmatrix} u \\ v \\ w \end{pmatrix} + \Gamma_1 \cdot \tilde{\Lambda} \cdot \begin{pmatrix} p \\ q \\ r \end{pmatrix}.
\end{aligned}
\tag{A.4.13}
$$

The rigid body state space equations of motion in the inertial frame are given by

$$
\dot{x}_{I_r} = A_r \cdot x_{I_r} + B_r \cdot u \tag{A.4.14}
$$

and the required state space equations in the body-fixed coordinate system are obtained by the following transformations

$$x_{I_r} = T_1 \cdot x_{B_r}$$

$$\begin{pmatrix} x \\ y \\ z \\ \phi \\ \theta \\ \psi \\ \dot{x} \\ \dot{y} \\ \dot{z} \\ \dot{\phi} \\ \dot{\theta} \\ \dot{\psi} \end{pmatrix} = T_1 \cdot \begin{pmatrix} \acute{x} \\ \acute{y} \\ \acute{z} \\ \acute{\phi} \\ \acute{\theta} \\ \acute{\psi} \\ u \\ v \\ w \\ p \\ q \\ r \end{pmatrix} \tag{A.4.15}$$

where the primed coordinates denote inertial displacement components of the aircraft center-of-gravity projected onto the body coordinate system, and the relevant matrices are given as

$$T_1 = \begin{bmatrix} \Gamma_1 & 0 & 0 & 0 \\ 0 & I & 0 & 0 \\ 0 & \tilde{\Lambda}_1 & \Gamma_1 & 0 \\ 0 & \tilde{\Lambda}_2 & 0 & \Gamma_2 \end{bmatrix} \tag{A.4.16}$$

with

$$\Gamma_2 = \begin{bmatrix} 1 & \sin\Phi_1 \tan\Theta_1 & \cos\Phi_1 \tan\Theta_1 \\ 0 & \cos\Phi_1 & -\sin\Phi_1 \\ 0 & \sin\Phi_1/\cos\Theta_1 & \cos\Phi_1/\cos\Theta_1 \end{bmatrix} \tag{A.4.17}$$

and

$$\tilde{\Lambda}_2 = \begin{bmatrix} \dot{\Theta}_1 \tan\Theta_1 & \dot{\Psi}_1/\cos\Theta_1 & 0 \\ \dot{\Psi}_1 \cos\Theta_1 & 0 & 0 \\ \dot{\Theta}_1/\cos\Theta_1 & \dot{\Psi}_1 \tan\Theta_1 & 0 \end{bmatrix}. \tag{A.4.18}$$

Similarly for state space velocities

$$\dot{x}_{I_r} = T_2 \cdot \dot{x}_{B_r} + T_3 \cdot x_{B_r} \tag{A.4.19}$$

where

$$T_2 = \begin{bmatrix} \Gamma_1 & 0 & 0 & 0 \\ 0 & \Gamma_2 & 0 & 0 \\ \Gamma_1 \tilde{\Omega} & \Gamma_1 \tilde{\Lambda} & \Gamma_1 & 0 \\ 0 & 0 & 0 & \Gamma_2 \end{bmatrix} \tag{A.4.20}$$

and

$$T_3 = \begin{bmatrix} 0 & \tilde{\Lambda}_1 & 0 & 0 \\ 0 & \tilde{\Lambda}_2 & 0 & 0 \\ 0 & 0 & 0 & 0 \\ 0 & 0 & 0 & 0 \end{bmatrix}. \tag{A.4.21}$$

No transformations are applied to elastic and aerodynamic state vectors. Rigid body state vector transformations are therefore given by

$$\begin{aligned} \dot{x}_{I_r} &= A x_{I_r} + Bu \\ T_2 \dot{x}_{B_r} &= A T_1 x_{B_r} - T_3 x_{B_r} + Bu \\ T_2 \dot{x}_{B_r} &= \left(A T_1 - T_3\right) x_{B_r} + Bu \end{aligned} \tag{A.4.22}$$

or

$$\begin{aligned} \dot{x}_{B_r} &= T_2^{-1} \left(A T_1 - T_3\right) x_{B_r} + T_2^{-1} Bu \\ &= A x_{B_r} + Bu. \end{aligned} \tag{A.4.23}$$

CHAPTER 5

AEROELASTIC AND AEROSERVOELASTIC MODELS

5.1 Aeroelastic Equation of Motion

Consider the generalized equation of motion that relates the structural dynamics and unsteady aerodynamics [56].

$$M\ddot{\eta} + C\dot{\eta} + K\eta + \overline{q}Q(s)\eta = 0 \tag{5.1}$$

For a system with n modes, define $M \in \mathbf{R}^{n \times n}$ as the mass matrix, $C \in \mathbf{R}^{n \times n}$ as the damping matrix and $K \in \mathbf{R}^{n \times n}$ as the stiffness matrix. $\overline{q} \in \mathbf{R}$ is a scalar representing the dynamic pressure and $Q(s) \in \mathbf{C}^{n \times n}$ is the matrix of unsteady aerodynamic forces. The aerodynamics are noted as a function of frequency but they also depend nonlinearly on Mach; therefore, a linear equation shown in (5.1) is required at each Mach number to describe the aircraft dynamics over a flight envelope.

Values of the aerodynamic force matrix at distinct frequencies can be derived using finite element models of the aircraft and panel methods for unsteady force calculations. An example of such a computer program that is developed for NASA is known as `STARS` [53, 54]. This code solves the subsonic aerodynamic equations using the doublet lattice method [14, 48]. The supersonic forces are generated using a different approach known as the constant panel method [4].

It is desired to formulate a linear time-invariant representation of the aerodynamic forces to incorporate them into the robust stability framework. Padé approximations can be used to compute a rational function approximation to the transfer function matrix.

$$Q(s) = A_0 + sA_1 + s^2A_2 + \frac{s}{s+\beta_1}A_3 + \frac{s}{s+\beta_2}A_4 \tag{5.2}$$

This form is often referred to as Roger's form [114]. The equation presented here only includes 2 lag terms, although more may be included. The poles of the lag terms, β_1 and β_2, are restricted to be real and positive to maintain system stability. The matrix elements of Roger's form may be computed using a least-square algorithm to fit the frequency-varying aerodynamic data.

The aerodynamic lag terms may be replaced in the formulation with a finite-dimensional state-space system represented by a transfer function matrix using Karpel's method [63].

$$Q(s) = A_0 + sA_1 + s^2A_2 + C(sI - A)^{-1}Bs \tag{5.3}$$

Standard system identification algorithms, including curve fitting or least squares approaches, may be used to compute the elements in the state-space portion of the formulation. It is assumed the A_i matrices are known from the low frequency aerodynamic force data or from experimental wind tunnel data.

A matrix fraction approach is also formulated to represent the aerodynamic forces as a linear time-invariant system [91, 92]. This generalized form computes rational matrix polynomials in a fractional form using a least-squares algorithm. It can be shown Roger's form and Karpel's form are subsets of the matrix fractional form.

The aeroelastic model can be formulated in the μ framework using any of these representations to describe the unsteady aerodynamics; however, a single form will be used for the sake of brevity to generalize the model development. Consider representing the aerodynamic force matrices as a finite-dimensional state-space system. This form is most similar to Karpel's form, excepting the additional A_i matrices are not explicitly accounted for in the formulation.

$$Q(s) = \begin{bmatrix} A_Q & B_Q \\ C_Q & D_Q \end{bmatrix} = D_Q + C_Q \left(sI - A_Q\right)^{-1} B_Q \tag{5.4}$$

Given the number of generalized states, n, and the number of aerodynamic states, n_Q, define $A_Q \in \mathbf{R}^{n_Q \times n_Q}, B_Q \in \mathbf{R}^{n_Q \times n}, C_Q \in \mathbf{R}^{n \times n_Q}$ and $D_Q \in \mathbf{R}^{n \times n}$ as the elements of the state-space system approximating $Q(s)$.

Fitting the aerodynamic data to a finite-dimensional state-space system is equivalent to fitting each term in the matrix to a real rational, proper transfer function. This seems to contradict the methods of Roger and Karpel, which form non-proper transfer functions due to the terms in s and s^2, but an approximation can demonstrate the relationship. Specifically, including a high frequency pole in the non-proper term, such as replacing s with $\frac{s}{s+10000}$, would not affect the low frequency region of interest but would ensure the resulting models are stable and proper. With this approximation, the forms of Roger and Karpel can be shown to be subsets of this method.

Standard frequency-domain system identification algorithms may generate a system with an arbitrarily large number of states. This state dimension does not greatly affect the computational cost of computing the robust stability analysis but extending the robustness analysis to controller synthesis places a greater emphasis on limiting the state dimension [94]. Low-order unsteady aerodynamic models can be generated using techniques such as model reduction algorithms [38]; however, these low-order models will still have the general form of (5.4).

It is straightforward to generate a state-space representation of the aeroelastic system including the state-space form of the unsteady aerodynamic forces. Consider the force vector, y, generated by the state vector, η, through the relationship $y = Qu$. The matrix equation for this relationship can be written using $x \in \mathbf{R}^{n_Q}$ as the vector of aerodynamic states.

$$\begin{bmatrix} \dot{x} \\ y \end{bmatrix} = \begin{bmatrix} A_Q & B_Q \\ C_Q & D_Q \end{bmatrix} \begin{bmatrix} x \\ \eta \end{bmatrix} \tag{5.5}$$

Formulate the aeroelastic differential equation utilizing x.

$$\begin{aligned} 0 &= M\ddot{\eta} + C\dot{\eta} + K\eta + \overline{q}Q(s)\eta \\ &= M\ddot{\eta} + C\dot{\eta} + K\eta + \overline{q}y \\ &= M\ddot{\eta} + C\dot{\eta} + K\eta + \overline{q}(C_Q x + D_Q \eta) \\ &= M\ddot{\eta} + C\dot{\eta} + (K + \overline{q}D_Q)\eta + \overline{q}C_Q x \end{aligned} \tag{5.6}$$

A state-space system is formulated utilizing the generalized states, η and $\dot{\eta}$, and the unsteady aerodynamic states, x. The state update matrix is determined by combining (5.5) and (5.6).

$$\begin{bmatrix} \dot{\eta} \\ \ddot{\eta} \\ \dot{x} \end{bmatrix} = \begin{bmatrix} 0 & I & 0 \\ -M^{-1}(K + \overline{q}D_Q) & -M^{-1}C & -\overline{q}M^{-1}C_Q \\ B_Q & 0 & A_Q \end{bmatrix} \begin{bmatrix} \eta \\ \dot{\eta} \\ x \end{bmatrix} \tag{5.7}$$

5.2 Nominal Aeroelastic Model

The generalized equation of motion for the nominal aeroelastic system can be expressed in a form suitable for using μ analysis to compute a flutter margin. The flutter margin is dependent on the flight condition parameters which result in a flutter instability while μ is defined to be the smallest perturbation among the set $\mathbf{\Delta}$ that causes an instability. The obvious approach to formulating flutter analysis in the μ framework is to introduce a perturbation to a flight condition parameter and find the smallest perturbation that causes an instability.

There are essentially two subsystems in the nominal aeroelastic model comprised of the structural dynamics involving mass, damping and stiffness matrices along with the unsteady aerodynamics scaled by the dynamic pressure. The generalized equation of motion at a constant Mach condition in (5.1) demonstrates the dynamic pressure linearly affects the dynamics at a constant Mach condition. Perturbations to dynamic pressure can thus enter the system through a feedback operator in a linear fractional manner which is perfectly suited to μ analysis.

Consider an additive perturbation, $\delta_{\overline{q}} \in \mathbf{R}$, that affects the nominal dynamic pressure, $\overline{q}_o$, to compute a value of $\overline{q}$.

$$\overline{q} = \overline{q}_o + \delta_{\overline{q}} \tag{5.8}$$

Separate terms in the system dynamics that involve $\delta_{\overline{q}}$.

$$\begin{aligned}
0 &= M\ddot{\eta} + C\dot{\eta} + (K + \overline{q}D_Q)\eta + \overline{q}C_Q x \\
&= M\ddot{\eta} + [C\dot{\eta} + (K + \overline{q}_o D_Q)\eta + \overline{q}_o C_Q x] + \delta_{\overline{q}}[D_Q\eta + C_Q x] \\
&= \ddot{\eta} + \left[M^{-1}C\dot{\eta} + M^{-1}(K + \overline{q}_o D_Q)\eta + \overline{q}_o M^{-1}C_Q x\right] \\
&\qquad\qquad + \delta_{\overline{q}}\left[M^{-1}D_Q\eta + M^{-1}C_Q x\right] \\
&= \ddot{\eta} + \left[M^{-1}C\dot{\eta} + M^{-1}(K + \overline{q}_o D_Q)\eta + \overline{q}_o M^{-1}C_Q x\right] + \delta_{\overline{q}} z \\
&= \ddot{\eta} + \left[M^{-1}C\dot{\eta} + M^{-1}(K + \overline{q}_o D_Q)\eta + \overline{q}_o M^{-1}C_Q x\right] + w
\end{aligned} \tag{5.9}$$

The signals z and w are introduced into this formulation to associate the perturbation on dynamic pressure to the nominal dynamics in a feedback manner. The signal z may be generated as an output of the plant since it is a linear combination of states.

$$z = M^{-1}D_Q\eta + M^{-1}C_Q x \tag{5.10}$$

The signal w is related to z by the dynamic pressure perturbation.

$$w = \delta_{\overline{q}} z \tag{5.11}$$

The state-space aeroelastic model for nominal stability analysis in the μ framework is formulated using the state update matrix determined by the dynamics at the nominal dynamic pressure and the additional input and output signals to introduce perturbations to the dynamic pressure. That perturbation $\delta_{\overline{q}}$ is not an explicit parameter in the state-space model since it only affects the plant through a feedback relationship as determined by the signals z and w. Define the transfer function $P(s)$ generated by state-space matrices such that $z = P(s)w$.

$$\begin{bmatrix} \dot{\eta} \\ \ddot{\eta} \\ \dot{x} \\ \hline z \end{bmatrix} = \left[\begin{array}{ccc|c} 0 & I & 0 & 0 \\ -M^{-1}(K + \overline{q}_o D_Q) & -M^{-1}C & -\overline{q}_o M^{-1}C_Q & -I \\ B_Q & 0 & A_Q & 0 \\ \hline M^{-1}D_Q & 0 & M^{-1}C_Q & 0 \end{array}\right] \begin{bmatrix} \eta \\ \dot{\eta} \\ x \\ \hline w \end{bmatrix} \tag{5.12}$$

The feedback interconnection between the perturbation in dynamic pressure and the nominal plant model parameterized around that perturbation is given in Figure 5.1. This interconnection is a linear fractional transformation and the small gain condition of Lemma 2.4.1 or the μ condition of Theorem 3.3.3 may be directly applied to analyze stability with respect to a variation in the flight condition caused by $\delta_{\overline{q}}$.

Formulating the nominal aeroelastic dynamics in the μ framework immediately demonstrates the procedure utilized in computing a flutter margin. Traditional flutter analysis algorithms such as the *p-k* method and the μ method as applied to Figure 5.1 are searching for a value of dynamic pressure that results in a flutter instability. The nominal flutter margin question may be posed which is answered by these methods.

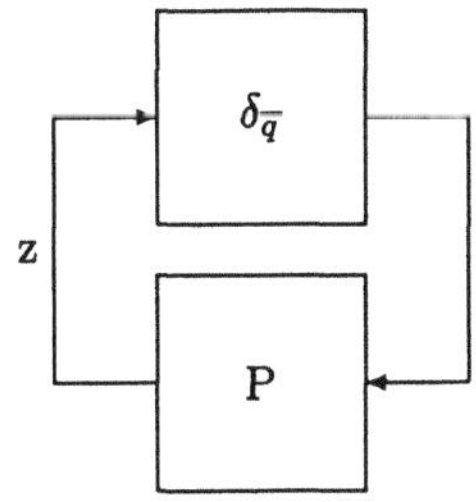

Fig. 5.1: Linear Fractional Transformation System for Nominal Stability Analysis in the μ Framework with Parameterization Around Perturbation in Dynamic Pressure

Question 5.2.1 *(nominal flutter margin) What is the largest perturbation to dynamic pressure for which the nominal aeroelastic dynamics are stable?*

The uncertainty block in Figure 5.1 is actually a repeated real scalar block whose dimension is the number of modes in the model. The number of free variables in the μ upper bound optimization, and consequently the computational cost of computing μ, is a function of the number of uncertainty parameters so nominal flutter margins are computed by optimizing over the single parameter of $\delta_{\overline{q}}$. In this way, the number of aerodynamic states, n_Q, does not directly affect the cost of the flutter estimation so low-order models are not preferred to high-order models. The only cost increase due to these additional states is computing the frequency response of the state-space matrix, which is generally much lower than the cost of computing μ.

It is straightforward to demonstrate the parameterization around dynamic pressure has not changed the nominal dynamics of the aeroelastic system. Simply compute the closed-loop transfer function with $\delta_{\overline{q}} = 0$ and the nominal system that matches (5.7) is recovered.

Wind tunnel and ground vibration testing can experimentally determine the aerodynamic stiffness and damping matrices which are more accurate than those approximated by finite element models. These matrices can be incorporated into a nominal model using the derivation in Appendix A.5.

It is important to note this procedure considers variations in dynamic pressure for an aeroelastic model at constant Mach number. The unsteady aerodynamics are highly nonlinear with variation in Mach number and attempts to model Mach variations in a linear fractional transformation may produce highly conservative flutter margins [51, 104]. The μ method as currently presented is considering flutter margins in terms of dynamic pressure as measured along lines of constant Mach. Flutter is a function of the two variables dynamic pressure and Mach so computing the dynamic pressure causing flutter for a dense set of discrete Mach values will generate an accurate portrait of the flutter margins.

5.3 Robust Aeroélastic Model

A robust aeroelastic model in the μ framework can be generated by associating uncertainty operators with the nominal model and including the parameterization around a perturbation in dynamic pressure. These uncertainty operators may resemble any of the forms presented in Section 3.2 including parametric uncertainty along with additive and multiplicative representations of dynamic uncertainty.

Choosing a reasonable uncertainty description is crucial for determining a valid robust flutter margin. This choice may arise logically from consideration of weaknesses in the modeling process, previous experience with aeroelastic analysis, and comparison with observed flight dynamics. A noncomprehensive examination of several obvious uncertainty descriptions that may be associated with an aeroelastic model is given in Chapter 6.

The linear fractional transformation is a valuable framework to utilize for formulating the robust aeroelastic model such that it is suitable for μ analysis. The various system blocks comprised of the nominal state-space model with associated uncertainties and any additional subsystem blocks and their associated uncertainties can be expressed as a single model and uncertainty operator.

Figure 5.2 presents the block structure used for μ analysis of the uncertain aeroelastic system. μ is computed with respect to a single block diagonal structured operator which contains the perturbation to dynamic pressure and the structured uncertainty operator along the diagonal. The perturbation to dynamic pressure is explicitly shown to distinguish $\delta_{\overline{q}}$ from the modeling uncertainty and emphasize it is a special operator used to describe a range of flight conditions.

The flutter margin computed for the uncertain system in Figure 5.2 is a more accurate margin than one computed with traditional methods such as *p-k* . These traditional methods address the nominal flutter problem in Question 5.2.1 while the robust flutter margin must consider the effect of the modeling uncertainty. The robust flutter margin actually finds the smallest perturbation to dynamic pressure for the entire set of plants formulated by the interconnection of the nominal dynamics and all elements $\Delta \in \boldsymbol{\Delta}$. Computing these margins is posed by Question 5.3.1.

Question 5.3.1 *(robust flutter margin) What is the largest perturbation to dynamic pressure for which the nominal aeroelastic dynamics are robustly stable to the entire range of modeling errors as described by the norm bounded set* $\boldsymbol{\Delta}$*?*

The introduction of uncertainty to robust aeroelastic model in Figure 5.2 does not significantly alter the block diagram as compared to the nominal aeroelastic model in Figure 5.1; however, the properties of the models are fundamentally different. Consider that the nominal flutter margins that answer

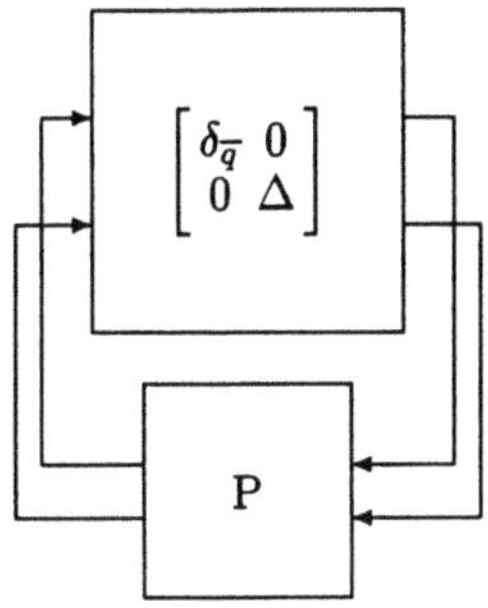

Fig. 5.2: Linear Fractional Transformation System for Robust Stability Analysis in the μ Framework with Parameterization Around Perturbation in Dynamic Pressure and Structured Uncertainty

Question 5.2.1 reflect the smallest destabilizing perturbation to the single nominal plant. Alternatively, the robust flutter margins that answer Question 5.3.1 actually reflect the smallest destabilizing perturbation to a set of plant models. This set of plant arises because the robust model is really the nominal dynamics and a set of norm-bounded operators related by feedback. Thus, the concept of a robust aeroelastic model may actually be misleading because a family of models is actually implied.

5.4 Robust Aeroservoelastic Model

Aeroservoelastic models must consider the interactions of aerodynamic, inertial, and structural forces that comprise the aeroelastic model along with the additional interactions resulting from including the control system dynamics with this aeroelastic model. Thus, these aeroservoelastic models have a strong and well-defined relationship to aeroelastic models.

A straightforward method to formulate aeroservoelastic models is to take advantage of the relationship with aeroelastic models by simply augmenting an existing aeroelastic model with a control system. The basic concept is to use linear fractional transformations and relate the open-loop model of Figure 5.2 with a control system through feedback relationships. In this way, nominal aeroservoelastic models are generated by relating nominal aeroelastic models with a control system while robust aeroservoelastic models are generated by relating robust aeroelastic models with a control system.

This method is a reasonable approach to consider because aeroelastic models are typically generated for flutter analysis before an aeroservoelastic analysis. Furthermore, even computational and finite-element approaches to generating aeroservoelastic models are usually based on an initial aeroelastic model. This method is even efficient when considering algebraic issues because the framework using linear fractional transformation is particularly convenient to use when relating several independent models.

An advantage to simply augmenting an existing aeroelastic model is that the resulting aeroservoelastic model automatically contains feedback signals to relate a perturbation to the dynamic pressure scaling the unsteady aerodynamics. This eliminates the need to repeat the derivation of state equations that are affected by feedback signals instead of explicitly depending on the perturbation.

Also, the control system may have its own associated uncertainties but it does not introduce any new uncertainties that affect the aeroelastic elements. Thus, the formulation of a robust aeroservoelastic model does not require a new evaluation of the structural dynamics and aerodynamics when generating an associated uncertainty description. The model that results from combine the aeroelastic and control models will retain the feedback interconnections relating the original uncertainty description with the proper aeroelastic elements because the linear fractional operations do not alter the structure of each element. This is especially important when the control system does not introduce any new uncertainties beyond the aeroelastic uncertainties so a robust aeroservoelastic model is relatively simple to formulate based on an existing robust aeroelastic model.

The additional elements that are associated with the control system that may not be contained in an aeroelastic model are usually represented by a feedback controller along with sensors and actuators. Modern controllers are implemented on digital computers and so their models have no uncertainty; however, the sensor and actuator models probably have noticeable uncertainties. The physical hardware for sensing and actuating always have nonlinearities and infinite-dimensional elements so linear models can not exactly represent the true dynamics.

The basic diagram of a robust aeroservoelastic model is shown in Figure 5.3. Beside the aeroelastic elements, this diagram uses A and Δ_A to represent the actuator model and its uncertainty, S and Δ_S to represent the sensor model and it uncertainty, and K and Δ_K to represent the controller and any uncertainty associated with the controller.

The basic properties of linear fractional transformations allow the complex unstructured model in Figure 5.3 to be written as an equivalent structured model similar to that diagram in Figure 5.2. Of course, the plant originally used in Figure 5.2 must be replaced with the closed-loop nominal plant and the uncertainty set must be replaced with a structured operator covering all the uncertainty elements but the essential appearance is the same.

Robust aeroservoelastic stability margins are computed by analyzing the smallest perturbation to dynamic pressure for which the closed-loop dynamics are not robustly stable to the modeling uncertainty. The similarity of

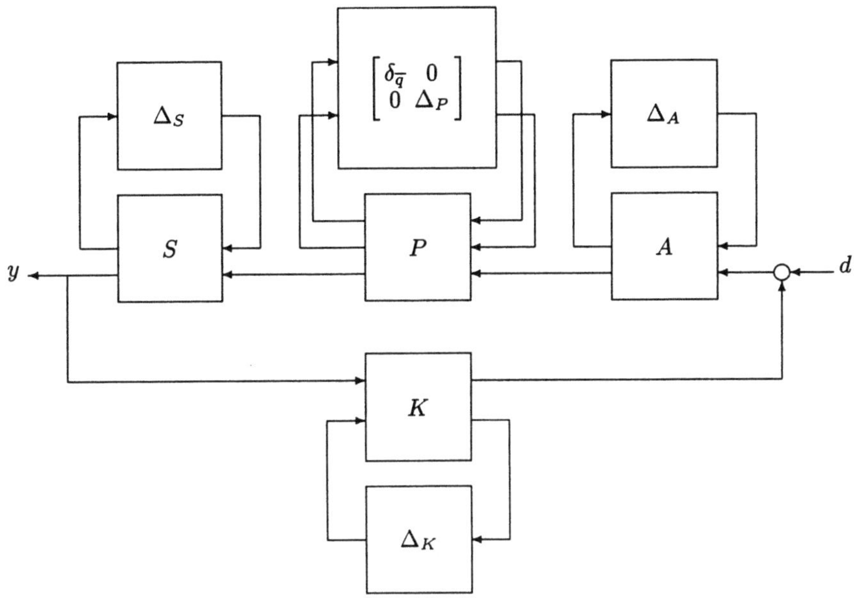

Fig. 5.3: Linear Fractional Transformation System for Robust Aeroservoelastic Stability Analysis in the μ Framework with Parameterization Around Perturbation in Dynamic Pressure and Structured Uncertainty

this concept to the concept of robust flutter margins is expected because of the relationship between the aeroelastic and aeroservoelastic models. Thus, the procedure used to compute robust aeroservoelastic models based on the closed-loop model is identical to the procedure used to compute robust flutter margins based on the open-loop model.

A potential problem with computing robust aeroservoelastic stability margins based on Figure 5.3 is the issue of controllers that gain scheduled as a function of dynamic pressure. This problem can be avoided by parameterizing the controller around dynamic pressure and augmenting the $\delta_{\bar{q}}$ block; however, this parameterization might not be possible for nonlinear controllers. Another possible solution is to simply ignore the gain-scheduled nature of the controller. The resulting margins would relate the closed-loop properties without any gain scheduling; however, these margins might be misleading if the controller is strongly dependent on dynamic pressure.

A.5 Models with Specific Aerodynamic Elements

State-space realizations of aeroelastic models can be formulated using representations of the unsteady aerodynamics that have no specialized structure. These models can also be generated for representations of the unsteady aerodynamics that do have some known structure. The resulting models can be parameterized around dynamic pressure and associated with an uncertainty description to allow for robust aeroelastic and aeroservoelastic stability margins to be computed.

The general formulation presented in Section 5.1 approximated the unsteady aerodynamic forces as a state-space system. This unstructured formulation does not explicitly include additional information about the aerodynamics that may be obtained from sources other than the computational model. In particular, the steady aerodynamics and first order derivatives are often measured from wind tunnel testing. Two matrix terms are introduced to the unsteady aerodynamic formulation to include this information.

$$\begin{aligned} Q(s) &= A_0 + sA_1 + \left[\begin{array}{c|c} A_Q & B_Q \\ \hline C_Q & D_Q \end{array}\right] \\ &= A_0 + sA_1 + D_Q + C_Q\left(sI - A_Q\right)^{-1} B_Q \end{aligned} \tag{A.5.1}$$

This system is quite similar to Karpel's form [63]. The main difference is the absence of an additional $s^2 A_2$ term. This term is typically not available from experimental data and is not explicitly represented in this formulation.

The state-space elements, $\{A_Q, B_Q, C_Q, D_Q\}$, are computed using standard system identification algorithms. The known matrix elements are subtracted from the unsteady aerodynamic force data and the resulting system, $Q_{ss}(s)$ is approximated by the state-space elements.

$$Q_{ss}(s) = Q(s) - A_0 - sA_1 = \left[\begin{array}{c|c} A_Q & B_Q \\ \hline C_Q & D_Q \end{array}\right] \tag{A.5.2}$$

Given the number of generalized states, n, such that $A_0, A_1 \in \mathbf{R}^{n\times n}$ and the number of aerodynamic states, n_Q, define $A_Q \in \mathbf{R}^{n_Q\times n_Q}, B_Q \in \mathbf{R}^{n_Q\times n}, C_Q \in \mathbf{R}^{n\times n_Q}$ and $D_Q \in \mathbf{R}^{n\times n}$ as the elements of the state-space system approximating $Q_{ss}(s)$.

The nominal aeroelastic model is extended to include the additional A_0 and A_1 terms. Consider the signal, y, generated by an input, η, to the state-space portion of force matrix through the relationship $y = Q_{ss}(s)\eta$. This relationship can be formulated in matrix form using the state-space matrices of Q_{ss} and $x \in \mathbf{R}^{n_Q}$ to represent the aerodynamic states.

$$\begin{bmatrix} \dot{x} \\ y \end{bmatrix} = \begin{bmatrix} A_Q & B_Q \\ C_Q & D_Q \end{bmatrix} \begin{bmatrix} x \\ \eta \end{bmatrix} \tag{A.5.3}$$

Formulate the aeroelastic differential equation utilizing x.

$$\begin{aligned}
0 &= M\ddot{\eta} + C\dot{\eta} + K\eta + \overline{q}Q(s)\eta \\
&= M\ddot{\eta} + C\dot{\eta} + K\eta \\
&\quad + \overline{q}\left(A_0 + sA_1 + D_Q + C_Q\left(sI - A_Q\right)^{-1} B_Q\right)\eta \\
&= M\ddot{\eta} + (C + \overline{q}A_1)\dot{\eta} + (K + \overline{q}A_0)\eta + \overline{q}Q_{ss}(s)\eta \\
&= M\ddot{\eta} + (C + \overline{q}A_1)\dot{\eta} + (K + \overline{q}A_0)\eta + \overline{q}y \\
&= M\ddot{\eta} + (C + \overline{q}A_1)\dot{\eta} + (K + \overline{q}A_0)\eta + \overline{q}(C_Q x + D_Q\eta) \\
&= M\ddot{\eta} + (C + \overline{q}A_1)\dot{\eta} + (K + \overline{q}A_0 + \overline{q}D_Q)\eta + \overline{q}C_Q x
\end{aligned} \tag{A.5.4}$$

A state-space system is formulated utilizing the generalized states, η and $\dot{\eta}$, and the unsteady aerodynamic states, x. The state update matrix is determined by the following 3 differential equations.

$$\begin{bmatrix} \dot{\eta} \\ \ddot{\eta} \\ \dot{x} \end{bmatrix} = \begin{bmatrix} 0 & I & 0 \\ -M^{-1}(K + \overline{q}A_0 + \overline{q}D_Q) & -M^{-1}(C + \overline{q}A_1) & -\overline{q}M^{-1}C_Q \\ B_Q & 0 & A_Q \end{bmatrix} \begin{bmatrix} \eta \\ \dot{\eta} \\ x \end{bmatrix}$$

This state-space aeroelastic model can be parameterized around dynamic pressure in the LFT framework and used for μ analysis. Consider an additive perturbation, $\delta_{\overline{q}} \in \mathbf{R}$, that affects the nominal dynamic pressure of $\overline{q}_o$.

$$\overline{q} = \overline{q}_o + \delta_{\overline{q}} \tag{A.5.5}$$

Separate terms in the system dynamics that involve $\delta_{\overline{q}}$.

$$\begin{aligned}
0 &= M\ddot{\eta} + (C + \overline{q}A_1)\dot{\eta} + (K + \overline{q}A_0 + \overline{q}D_Q)\eta + \overline{q}C_Q x \\
&= M\ddot{\eta} + (C + \overline{q}_o A_1)\dot{\eta} + (K + \overline{q}_o A_0 + \overline{q}_o D_Q)\eta + \overline{q}_o C_Q x \\
&\quad + \delta_{\overline{q}}\left[A_1\dot{\eta} + A_0\eta + D_Q\eta + C_Q x\right] \\
&= \ddot{\eta} + M^{-1}(C + \overline{q}_o A_1)\dot{\eta} + M^{-1}(K + \overline{q}_o A_0 + \overline{q}_o D_Q)\eta \\
&\quad + \overline{q}_o M^{-1} C_Q x \\
&\quad + \delta_{\overline{q}}\left[M^{-1}A_1\dot{\eta} + M^{-1}A_0\eta + M^{-1}D_Q\eta + M^{-1}C_Q x\right] \\
&= \ddot{\eta} + M^{-1}(C + \overline{q}_o A_1)\dot{\eta} + M^{-1}(K + \overline{q}_o A_0 + \overline{q}_o D_Q)\eta \\
&\quad + \overline{q}_o M^{-1} C_Q x + \delta_{\overline{q}} z \\
&= \ddot{\eta} + M^{-1}(C + \overline{q}_o A_1)\dot{\eta} + M^{-1}(K + \overline{q}_o A_0 + \overline{q}_o D_Q)\eta \\
&\quad + \overline{q}_o M^{-1} C_Q x + w
\end{aligned} \tag{A.5.6}$$

The signals z and w are introduced into this formulation to associate the perturbation in dynamic pressure to the nominal dynamics in a feedback manner. The signal z may be generated as an output of the plant since it is a linear combination of states.

$$z = M^{-1}A_1\dot{\eta} + M^{-1}A_0\eta + M^{-1}D_Q\eta + M^{-1}C_Q x \tag{A.5.7}$$

The signal w is related to z by the dynamic pressure perturbation.

$$w = \delta_{\overline{q}} z \tag{A.5.8}$$

The state-space aeroelastic model for nominal stability analysis in the μ framework is formulated using the state update matrix determined by the dynamics at the nominal dynamic pressure and the additional input and output signals to introduce perturbations to the dynamic pressure. That perturbation $\delta_{\overline{q}}$ is not an explicit parameter in the state-space model since it only affects the plant through a feedback relationship as determined by the signals z and w. Define the transfer function $P(s)$ generated by state-space matrices such that $z = P(s)w$.

$$\begin{bmatrix} \dot{\eta} \\ \ddot{\eta} \\ \dot{x} \\ \hline z \end{bmatrix} = \left[\begin{array}{ccc|c} & & & 0 \\ & A & & -I \\ & & & 0 \\ \hline M^{-1}(A_0 + D_Q) & M^{-1}A_1 & M^{-1}C_Q & 0 \end{array}\right] \begin{bmatrix} \eta \\ \dot{\eta} \\ x \\ \hline w \end{bmatrix} \tag{A.5.9}$$

where

$$A = \begin{bmatrix} 0 & I & 0 \\ -M^{-1}(K + \overline{q}_o A_0 + \overline{q}_o D_Q) & -M^{-1}(C + \overline{q}_o A_1) & -\overline{q}_o M^{-1} C_Q \\ B_Q & 0 & A_Q \end{bmatrix}$$

The feedback interconnection between the perturbation in dynamic pressure and the nominal plant model parameterized around that perturbation is given in Figure 5.1. This interconnection is a linear fractional transformation and the small gain condition of Lemma 2.4.1 or the μ condition of Theorem 3.3.3 may be directly applied to analyze stability with respect to a variation in the flight condition $\overline{q}$. It is seen this system reduces to the parameterized model presented in Section 5.2 if $A_0 = A_1 = 0$.

CHAPTER 6
UNCERTAINTY DESCRIPTIONS

6.1 Parametric Uncertainty in Structural Models

Parametric uncertainty denotes operators which describe errors and unmodeled variations in specific elements and coefficients in dynamic system equations. Recall the generalized aeroelastic equation of motion for state vector $\eta \in \mathbf{R}^n$ [56].

$$M\ddot{\eta} + C\dot{\eta} + K\eta + \bar{q}Q(s)\eta = 0 \tag{6.1}$$

Robust stability margins computed with the μ method are strongly affected by the choice of uncertainty descriptions associated with these dynamics so this uncertainty must be a reasonable indicator of potential modeling errors. Parametric uncertainty may be directly associated with the structural matrices to indicate specific errors in the finite element model.

Define an operator, $\Delta_K \in \mathbf{R}^{n\times n}$, which describes additive uncertainty of a nominal stiffness matrix K_o. Associate a weighting matrix, $W_K \in \mathbf{R}^{n\times n}$, with this uncertainty such that a stability analysis should consider a range of stiffness matrices described by all Δ_K with $\|\Delta_K\|_\infty \leq 1$.

$$K = K_o + W_K \Delta_K \tag{6.2}$$

Parametric uncertainty can also be associated with structural elements in a multiplicative relationship. Define an operator, $\Delta_C \in \mathbf{R}^{n\times n}$, which describes multiplicative uncertainty of the nominal damping matrix C_o. A weighting $W_C \in \mathbf{R}^{n\times n}$ is again associated with the uncertainty such that the anticipated range of damping matrices for robust stability analysis is described by all Δ_C with $\|\Delta_C\|_\infty \leq 1$.

$$C = C_o \left(I + W_C \Delta_C\right) \tag{6.3}$$

The choice of additive uncertainty for Δ_K and multiplicative uncertainty for Δ_C does not reflect any generalized assumptions about the proper way to model errors in stiffness and damping; rather, each type is included to demonstrate the different mathematical derivations. Additive and multiplicative operators are common types of uncertainty models so it is instructive to demonstrate how these types of uncertainty are associated with a structural model. The actual choice of Δ_K and Δ_C is problem dependent and may vary with different aircraft and different finite element modeling procedures.

The uncertainty operators are described in this Section as constant matrices, $\Delta_K, \Delta_C \in \mathbf{R}^{n\times n}$, with real elements since they describe perturbations to the generalized stiffness and damping matrices which are usually real. These operators are often additionally constrained to be diagonal operators with n independent parameters since the generalized stiffness and damping are often computed as real diagonal matrices. The real and diagonal nature of these uncertainties is not required for μ analysis so full block complex uncertainties may be used if they better describe the nature of the modeling errors.

Also, the weighting functions $W_K, W_C \in \mathbf{R}^{n\times n}$ are presented as constant and real matrices since they are associated with constant and real stiffness and damping matrices. These constraints on the weightings may be relaxed if the nature of the uncertainty is best described by complex and frequency-varying weighting functions; however, a complex frequency-varying weighting and uncertainty are generally not used to describe modeling errors of real parameters.

Substitute the uncertain K and C into the equation of motion including the state-space representation of the unsteady aerodynamic forces $Q(s)$ presented in Section 5.1. Introduce a perturbation $\delta_{\overline{q}}$ to dynamic pressure and separate the nominal dynamics from the unknown terms.

$$\begin{aligned}
0 &= M\ddot{\eta} + C\dot{\eta} + (K + \overline{q}D_Q)\eta + \overline{q}C_Q x \\
&= M\ddot{\eta} + C_o\left(I + W_C\Delta_C\right)\dot{\eta} + \left(K_o + W_K\Delta_K + \left(\overline{q}_o + \delta_{\overline{q}}\right) D_Q\right)\eta \\
&\quad + \left(\overline{q}_o + \delta_{\overline{q}}\right) C_Q x \\
&= \ddot{\eta} + \left[M^{-1}C_o\dot{\eta} + M^{-1}(K_o + \overline{q}_o D_Q)\eta + \overline{q}_o M^{-1}C_Q x\right] \\
&\quad + \delta_{\overline{q}}\left[M^{-1}D_Q\eta + M^{-1}C_Q x\right] \\
&\quad + \Delta_K\left[M^{-1}W_K\eta\right] + \Delta_C\left[M^{-1}C_o W_C\dot{\eta}\right] \\
&= \ddot{\eta} + \left[M^{-1}C_o\dot{\eta} + M^{-1}(K_o + \overline{q}_o D_Q)\eta + \overline{q}_o M^{-1}C_Q x\right] \\
&\quad + \delta_{\overline{q}} z_{\overline{q}} + \Delta_K z_K + \Delta_C z_C \\
&= \ddot{\eta} + \left[M^{-1}C_o\dot{\eta} + M^{-1}(K_o + \overline{q}_o D_Q)\eta + \overline{q}_o M^{-1}C_Q x\right] \\
&\quad + w_{\overline{q}} + w_K + w_C
\end{aligned} \tag{6.4}$$

The signals $z_{\overline{q}}$ and $w_{\overline{q}}$ are introduced in Section 5.2 to relate the perturbation in dynamic pressure to the nominal plant through a feedback relationship. The feedback operation $w_{\overline{q}} = \delta_{\overline{q}} z_{\overline{q}}$ is used where $z_{\overline{q}}$ is a linear combination of the states of the plant.

$$z_{\overline{q}} = M^{-1}D_Q\eta + M^{-1}C_Q x \tag{6.5}$$

Additional signals are introduced to the plant formulation with z_K and w_K associated with the uncertainty in the stiffness matrix and z_C and w_C associated with the uncertainty in the damping matrix. The outputs of the plant z_K and z_C are formulated as linear combinations of the states.

$$z_K = M^{-1} W_K \eta \tag{6.6}$$
$$z_C = M^{-1} W_C C_o \dot{\eta} \tag{6.7}$$

The feedback mechanism to describe the modeling uncertainty uses a relationship between these output signals and the w_K and w_C input signals.

$$w_K = \Delta_K z_K \tag{6.8}$$
$$w_C = \Delta_C z_C \tag{6.9}$$

The state-space plant matrix can be formulated using these additional input and output signals.

$$\begin{bmatrix} \dot{\eta} \\ \ddot{\eta} \\ \dot{x} \\ \hline z_{\overline{q}} \\ z_K \\ z_C \end{bmatrix} = \left[\begin{array}{ccc|ccc} 0 & I & 0 & 0 & 0 & 0 \\ -M^{-1}(K_o + \overline{q}_o D_Q) & -M^{-1} C_o & -\overline{q}_o M^{-1} C_Q & -I & -I & -I \\ B_Q & 0 & A_Q & 0 & 0 & 0 \\ \hline M^{-1} D_Q & 0 & M^{-1} C_Q & 0 & 0 & 0 \\ M^{-1} W_K & 0 & 0 & 0 & 0 & 0 \\ 0 & M^{-1} C_o W_C & 0 & 0 & 0 & 0 \end{array}\right] \begin{bmatrix} \eta \\ \dot{\eta} \\ x \\ \hline w_{\overline{q}} \\ w_K \\ w_C \end{bmatrix} \tag{6.10}$$

The uncertainty operators and perturbation to dynamic pressure affect this plant formulation in a feedback manner as shown in Figure 6.1.

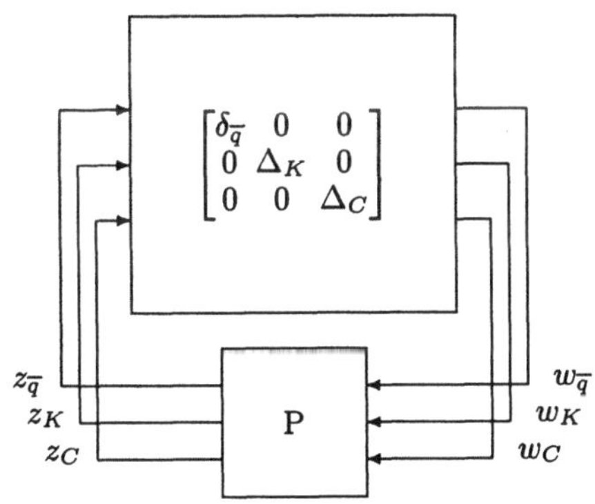

Fig. 6.1: Linear Fractional Transformation System for Robust Stability Analysis in the μ Framework with Parameterization Around Perturbation in Dynamic Pressure and Uncertainty in Structural Stiffness and Damping Matrices

The formulation does not directly allow uncertainty in mass to be described by a feedback operator. The LFT and μ frameworks require uncertainty operators to affect the nominal dynamics in a linear manner and this precludes introducing mass uncertainty. The inverse of the mass matrix scales most terms in the state matrix of $P(s)$ including terms involving $\overline{q}$. Associating a mass uncertainty operator Δ_M with the mass matrix scaling $\overline{q}$ would introduce terms of $\Delta_M \delta_{\overline{q}}$ which is a nonlinear function of uncertainty operators and can not be directly considered by the μ method.

6.2 Parametric Uncertainty in Aerodynamic Models

The unsteady aerodynamic forces $Q(s) \in \mathbf{C}^{n \times n}$ can be represented as a state-space model with n_Q states.

$$Q(s) = D_Q + C_Q \left(sI - A_Q\right)^{-1} B_Q \tag{6.11}$$

Parametric uncertainty can be associated with the matrix elements of this state-space representation to describe errors. These errors may result from several sources in the modeling procedure including computational fluid dynamic algorithms which determine the frequency-varying forces and the system identification algorithms which represent the computational values as a state-space system.

Define an operator $\Delta_{A_Q} \in \mathbf{R}^{n_Q \times n_Q}$ to describe uncertainty in the state matrix of $Q(s)$. This operator directly affects a nominal A_{Q_o} and describes errors and variations in the poles of the state-space representation of the unsteady aerodynamic forces. Include a weighting function $W_{A_Q} \in \mathbf{R}^{n \times n}$ such that the range of state matrices to be considered by robust stability analysis is described by all Δ_{A_Q} with $||\Delta_{A_Q}||_\infty \leq 1$.

$$A_Q = A_{Q_o} + W_{A_Q} \Delta_{A_Q} \tag{6.12}$$

Define also an operator $\Delta_{B_Q} \in \mathbf{R}^{n_Q \times n}$ to describe multiplicative uncertainty in a nominal B_{Q_o} matrix of $Q(s)$. A weighting function $W_{B_Q} \in \mathbf{R}^{n_Q \times n}$ is associated with this uncertainty such that the range of possible B_Q matrices is described by all Δ_{B_Q} with $||\Delta_{B_Q}||_\infty \leq 1$.

$$B_Q = B_{Q_o} \left(I + W_{B_Q} \Delta_{B_Q}\right) \tag{6.13}$$

The choice of additive and multiplicative operators is made for reasons similar to those presented in Section 6.1. One of each type of uncertainty is included to demonstrate the derivation procedures of how each uncertainty is associated with the nominal aeroelastic dynamics in a feedback relationship. The actual choice of which type of uncertainty is most suitable to describe errors in A_Q and B_Q is problem dependent.

Also, defining the uncertainty operators to be real and defining weightings which are real and constant is not a requirement. This Section presents the problem formulation with these definitions since it may make intuitive sense to associate these types of uncertainties and weightings with the constant real A_Q and B_Q matrices but certainly μ can compute robust flutter margins with respect to complex frequency-varying weighted uncertainties also.

The nominal aeroelastic model in Section 5.2 defined the vector x as the states associated with the state-space model $Q(s)$. The matrices A_Q and B_Q directly affect the aeroelastic dynamics only through the state derivative equation for x and do not appear in the other state derivative equations. The aeroelastic plant can be formulated in the μ framework by substituting the uncertain values of A_Q and B_Q into this state derivative equation without considering changes in the derivative equations for the remaining states η and $\dot{\eta}$.

$$\begin{aligned}
\dot{x} &= A_Q x + B_Q \eta \\
&= \left(A_{Q_o} + W_{A_Q}\Delta_{A_Q}\right) x + B_{Q_o}\left(I + W_{B_Q}\Delta_{B_Q}\right)\eta \\
&= A_{Q_o} x + B_{Q_o}\eta + \Delta_{A_Q} W_{A_Q} x + \Delta_{B_Q} W_{B_Q} B_{Q_o}\eta \\
&= A_{Q_o} x + B_{Q_o}\eta + \Delta_{A_Q} z_{A_Q} + \Delta_{B_Q} z_{B_Q} \\
&= A_{Q_o} x + B_{Q_o}\eta + w_{A_Q} + w_{B_Q}
\end{aligned} \tag{6.14}$$

Several signals are introduced to this equation with z_{A_Q} and w_{A_Q} associated with the uncertainty in A_Q and z_{B_Q} and w_{B_Q} associated with the uncertainty in B_Q. The signals z_{A_Q} and z_{B_Q} are output from the plant matrix as linear combinations of the states.

$$z_{A_Q} = W_{A_Q} x \tag{6.15}$$

$$z_{B_Q} = W_{B_Q} B_{Q_o} \eta \tag{6.16}$$

The feedback mechanism to describe the modeling uncertainty uses a relationship between these output signals and the w_K and w_C input signals.

$$w_{A_Q} = \Delta_{A_Q} z_{A_Q} \tag{6.17}$$

$$w_{B_Q} = \Delta_{B_Q} z_{B_Q} \tag{6.18}$$

The state-space plant matrix can be formulated using these additional input and output signals.

$$\begin{bmatrix} \dot{\eta} \\ \ddot{\eta} \\ \dot{x} \\ \hline z_{\bar{q}} \\ z_{A_Q} \\ z_{B_Q} \end{bmatrix} = \left[\begin{array}{ccc|ccc} 0 & I & 0 & 0 & 0 & 0 \\ -M^{-1}(K_o + \bar{q}_o D_Q) & -M^{-1}C_o & -\bar{q}_o M^{-1} C_Q & -I & 0 & 0 \\ B_{Q_o} & 0 & A_{Q_o} & 0 & I & I \\ \hline M^{-1}D_Q & 0 & M^{-1}C_Q & 0 & 0 & 0 \\ 0 & 0 & W_{A_Q} & 0 & 0 & 0 \\ W_{B_Q}B_{Q_o} & 0 & 0 & 0 & 0 & 0 \end{array}\right] \begin{bmatrix} \eta \\ \dot{\eta} \\ x \\ \hline w_{\bar{q}} \\ w_{A_Q} \\ w_{B_Q} \end{bmatrix} \tag{6.19}$$

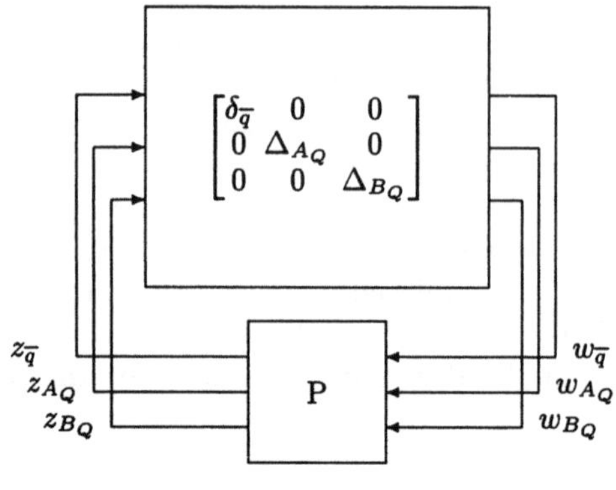

Fig. 6.2: Linear Fractional Transformation System for Robust Stability Analysis in the μ Framework with Parameterization Around Perturbation in Dynamic Pressure and Uncertainty in A_Q and B_Q Matrices of the State-Space $Q(s)$ Model

The uncertainty operators and perturbation to dynamic pressure affect this plant formulation in a feedback manner as shown in Figure 6.2.

Associating a Δ_{B_Q} uncertainty operator with the B_Q matrix may not seem immediately useful since it is often more intuitive to consider errors in the poles determined by the A_Q matrix. This Δ_{B_Q} uncertainty may be essential to accurately describe the modeling errors since there may be errors in terms common to both A_Q and B_Q. Such a situation arises for certain modeling representations of aerodynamic lags. Consider a simplified Roger's form $Q(s)$ matrix which uses two Padé approximations to represent lag terms.

$$Q(s) = \frac{s}{s+\beta_1} + \frac{s}{s+\beta_2} = \left[\begin{array}{cc|c} -\beta_1 & 0 & -\beta_1 \\ 0 & -\beta_2 & -\beta_2 \\ \hline 1 & 1 & 2 \end{array}\right] \tag{6.20}$$

The poles of this system are determined entirely by the A_Q matrix so uncertainty in the poles can be entirely described by a Δ_{A_Q} operator associated with the A_Q matrix. A similar uncertainty Δ_{B_Q} should also be associated with the B_Q matrix in this case because the β_1 and β_2 terms appear in both A_Q and B_Q. Allowing variation in A_Q but not in B_Q will introduce unwanted zeros to the system so the proper way to model pole uncertainty for this formulation is to include both Δ_{A_Q} and Δ_{B_Q} operators.

The Padé approximation appears often in aeroelastic models so it may be useful to demonstrate the LFT formulation of $Q(s)$ which includes the uncertainties in poles β_1 and β_2 in a feedback relationship. The uncertainties in $Q(s)$ can be developed distinctly from the structural uncertainties since LFT operations allow the structural and aerodynamic models to be combined into a single plant model with a structured uncertainty description [8].

Define real scalar operators Δ_{β_1} and Δ_{β_2} to describe uncertainty associated with nominal values of the poles β_{1_o} and β_{2_o}. Real scalar weightings W_{β_1} and W_{β_2} normalize the uncertainty such that the range of poles to be considered by robust stability analysis is described by all Δ_{β_1} with $||\Delta_{\beta_1}||_\infty \leq 1$ and all Δ_{β_2} with $||\Delta_{\beta_2}||_\infty \leq 1$.

$$\beta_1 = \beta_{1_o} + W_{\beta_1}\Delta_{\beta_1} \tag{6.21}$$
$$\beta_2 = \beta_{2_o} + W_{\beta_2}\Delta_{\beta_2} \tag{6.22}$$

Define states x_1 and x_2 of $Q(s)$ and consider an input signal u_Q which generates the output signal y_Q by the relationship $y_Q = Qu_Q$. Substitute the uncertain β_1 into the state equation of x_1.

$$\begin{aligned}
\dot{x}_1 &= -\beta_1 x_1 - \beta_1 u_Q \\
&= -\left(\beta_{1_o} + W_{\beta_1}\Delta_{\beta_1}\right)x_1 - \left(\beta_{1_o} + W_{\beta_1}\Delta_{\beta_1}\right)u_Q \\
&= -\beta_{1_o}x_1 - \beta_{1_o}u_Q - \Delta_{\beta_1}\left(W_{\beta_1}x_1 + W_{\beta_1}u_Q\right) \\
&= -\beta_{1_o}x_1 - \beta_{1_o}u_Q - \Delta_{\beta_1}z_{\beta_1} \\
&= -\beta_{1_o}x_1 - \beta_{1_o}u_Q - w_{\beta_1}
\end{aligned} \tag{6.23}$$

Perform a similar derivation for the state equation of x_2.

$$\begin{aligned}
\dot{x}_2 &= -\beta_2 x_2 - \beta_2 u_Q \\
&= -\left(\beta_{2_o} + W_{\beta_2}\Delta_{\beta_2}\right)x_2 - \left(\beta_{2_o} + W_{\beta_2}\Delta_{\beta_2}\right)u_Q \\
&= -\beta_{2_o}x_2 - \beta_{2_o}u_Q - \Delta_{\beta_2}\left(W_{\beta_2}x_2 + W_{\beta_2}u_Q\right) \\
&= -\beta_{2_o}x_2 - \beta_{2_o}u_Q - \Delta_{\beta_2}z_{\beta_2} \\
&= -\beta_{2_o}x_2 - \beta_{2_o}u_Q - w_{\beta_2}
\end{aligned} \tag{6.24}$$

The signals z_{β_1} and w_{β_1} are introduced as plant output and input signals to relate the uncertainty Δ_{β_1} to the nominal dynamics through the feedback of $w_{\beta_1} = \Delta_{\beta_1}z_{\beta_1}$. The signals z_{β_1} and w_{β_1} are similarly introduced to relate the uncertainty Δ_{β_2} to the nominal dynamics in a feedback manner through $w_{\beta_2} = \Delta_{\beta_2}z_{\beta_2}$. The state-space matrix can be formulated to describe the nominal $Q_o(s)$ with these additional input and output signals.

$$\left[\begin{array}{c} \dot{x}_1 \\ \dot{x}_2 \\ \hline z_{\beta_1} \\ z_{\beta_2} \\ y_Q \end{array}\right] = \left[\begin{array}{cc|ccc} -\beta_{1_o} & 0 & -1 & 0 & -\beta_{1_o} \\ 0 & -\beta_{2_o} & 0 & -1 & -\beta_{2_o} \\ \hline W_{\beta_1} & 0 & 0 & 0 & W_{\beta_1} \\ 0 & W_{\beta_2} & 0 & 0 & W_{\beta_2} \\ 1 & 1 & 0 & 0 & 2 \end{array}\right] \left[\begin{array}{c} x_1 \\ x_2 \\ \hline w_{\beta_1} \\ w_{\beta_2} \\ u_Q \end{array}\right] \tag{6.25}$$

The block diagram for $Q(s)$ formulated by relating the nominal model with its associated uncertainty operators is shown in Figure 6.3.

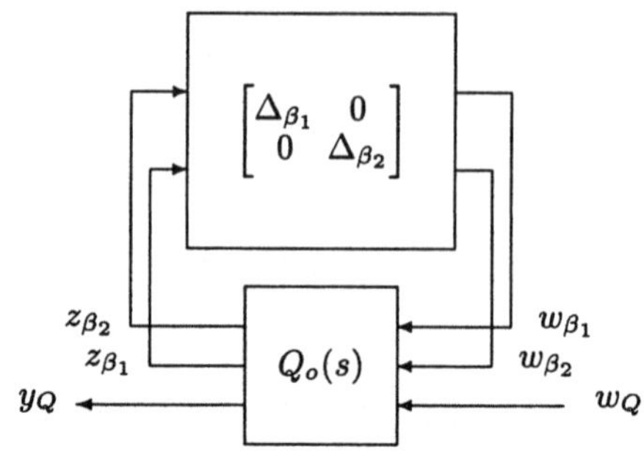

Fig. 6.3: Linear Fractional Transformation System Describing Padé Approximation to Represent Unsteady Aerodynamic Force Matrix in the μ Framework with Uncertainty in Lag Terms

6.3 Parametric Uncertainty in Modal Models

Structural models used for aeroelastic and aeroservoelastic analysis are often formulated in modal form such that elements of the state matrix describe properties associated with modal properties of the dynamics. These elements may represent real and imaginary components of the eigenvalues or natural frequencies and dampings of the modes. The modal elements are formulated by combining terms in the equations of motion, such as mass and stiffness or damping parameters, to simplify the formulation. Thus, it may be difficult to associate parametric uncertainty with a modal model to describe potential modeling errors with a specific structural parameter in an equation of motion.

Parametric uncertainty may be associated with a modal model to describe uncertainties that originate with physical parameters in the equation of motion but are observed through the modal properties of the dynamics. This type of uncertainty notes that errors in the equations of motion may be described by resulting errors in the natural frequencies and dampings in the transfer functions of the dynamics. For example, there is often a direct relationship between stiffness and natural frequency for a simple system through $\omega = \sqrt{k/m}$ so an uncertainty can be associated with either ω or k and the effect is identical.

Unfortunately, the uncertainty description for modal models is sometimes not realizable as a linear fractional transformation between the nominal modal model and a feedback uncertainty. Consider a single-mode system that generates the measurement y in response to an input u.

$$y = \frac{1}{s^2 + 2\zeta\omega s + w^2} u \tag{6.26}$$

A state-space realization of this system can be formulated using $\eta = y$ and $\dot{\eta} = \dot{y}$ as the states.

$$\begin{bmatrix} \dot{\eta} \\ \ddot{\eta} \\ \hline y \end{bmatrix} = \left[\begin{array}{cc|c} 0 & 1 & 0 \\ -\omega^2 & 2\zeta\omega & 1 \\ \hline 1 & 0 & 0 \end{array}\right] \begin{bmatrix} \eta \\ \dot{\eta} \\ \hline u \end{bmatrix} \tag{6.27}$$

Assume there are potential errors in the natural frequency and damping parameters. Associate an uncertainty operator, $\Delta_\zeta \in \mathbf{R}$, with the nominal damping, ζ_o, such that the range of damping values to considered for stability analysis is described by all $\zeta = \zeta_o + \Delta_\zeta$ with $\|\Delta_\zeta\|_\infty \leq 1$. Similarly, define $\Delta_\omega \in \mathbf{R}$ to associate with the nominal natural frequency, Δ_ω, such that true natural frequency of the system is assumed to lie within the set described by $\omega = \omega_o + \Delta_\omega$ with $\|\Delta_\omega\|_\infty \leq 1$.

$$\zeta = \zeta_o + \Delta_\zeta \tag{6.28}$$
$$\omega = \omega_o + \Delta_\omega \tag{6.29}$$

Introduce these uncertain parameters into the state equation for $\ddot{\eta}$.

$$\begin{aligned} \ddot{\eta} &= -\omega^2\eta - 2\zeta\omega\dot{\eta} + u \\ &= -\left(\omega_o + \Delta_\omega\right)^2\eta - 2\left(\omega_o + \Delta_\zeta\right)\left(\omega_o + \Delta_\omega\right)\dot{\eta} + u \\ &= -\omega_o^2\eta - 2\zeta_o\omega_o\dot{\eta} + u - \Delta_\omega 2\omega_o\eta - \Delta_\zeta 2\omega_o\eta - \Delta_\omega 2\zeta_o\dot{\eta} \\ &\quad -\Delta_\omega^2\eta - \Delta_\zeta\Delta_\omega 2\dot{\eta} \end{aligned} \tag{6.30}$$

The μ method is unable to accurately compute a measure of robustness for the system with state equation described by (6.30) because of the terms involving Δ_ω^2 and $\Delta_\zeta\Delta_\omega$. The structured singular value is currently defined only to measure stability for interconnections of linear operators and so the presence of these nonlinear uncertainty terms precludes the proper application of μ. Thus, different uncertainty descriptions must be developed to represent parametric uncertainty in a modal model.

One approximation to an uncertainty description that will account for the uncertainties in (6.28) and (6.29) is simply to ignore the nonlinear terms in (6.30). This approximation may be fine if the stability is most affected by small values of uncertainty and not affected when Δ_ω or Δ_ζ are near unity. Input and output signals are defined to relate the nominal plant to the uncertainty approximation through the relationship $w_\omega = \Delta_\omega z_\omega$ and $w_\zeta = \Delta_\zeta z_\zeta$ where the output signals z_ω and z_ζ are computed as functions of the plant states.

$$z_\zeta = -2\omega_o\eta \tag{6.31}$$
$$z_\omega = -2\omega_o\eta - 2\zeta_o\dot{\eta} \tag{6.32}$$

A state-space realization of the robust model can then be formulated directly from the linear terms of (6.30).

$$\begin{bmatrix} \dot{\eta} \\ \ddot{\eta} \\ \hline z_\zeta \\ z_\omega \\ y \end{bmatrix} = \left[\begin{array}{cc|c} 0 & 1 & 0\,0\,0 \\ -\omega_o^2 & -2\zeta_o\omega_o & 1\,1\,1 \\ \hline -2\omega_o & 0 & 0\,0\,0 \\ -2\omega_o & -2\zeta_o & 0\,0\,0 \\ 1 & 0 & 0\,0\,0 \end{array}\right] \begin{bmatrix} \eta \\ \dot{\eta} \\ \hline w_\zeta \\ w_\omega \\ u \end{bmatrix} \tag{6.33}$$

The nominal plant P that is described by the state-space realization of (6.33) and the associated the uncertainty operators Δ_ζ and Δ_ω are related through the feedback relationship shown in Figure 6.4.

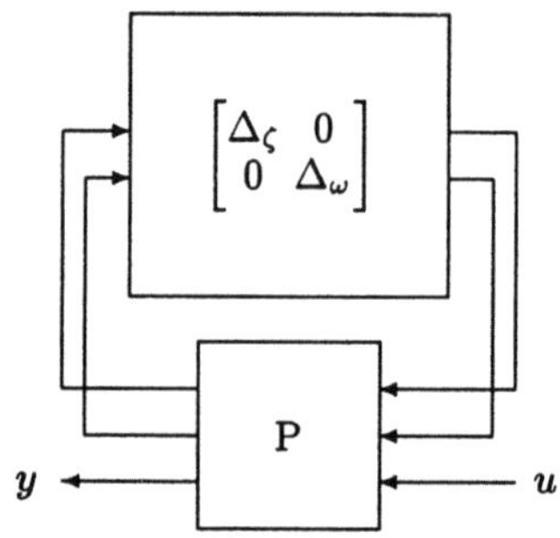

Fig. 6.4: Nominal Plant and Linear Approximation to Parametric Uncertainty in Damping and Natural Frequency

Another formulation that approximates parametric uncertainty in both natural frequency and damping elements of modal models is derived by considering errors in damping and the square of natural frequency. This formulation is an approximation because it only represents errors in ω^2 but ignores the corresponding error ω; however, it may be a useful approximation that is straightforward to implement.

Consider the equation for damping and squared natural frequency that results by associating operators $\delta_\zeta \in \mathbf{R}$ and $\delta_\omega \in \mathbf{R}$ with nominal elements ζ_o and ω_o^2. Also, introduce weightings $W_\zeta \in \mathbf{R}$ and $W_\omega \in \mathbf{R}$ such that the anticipated modeling errors are described by all δ_ζ with $||\delta_\zeta||_\infty \leq 1$ and all δ_ω with $||\delta_\omega||_\infty \leq 1$.

$$\zeta = \zeta_o + W_\zeta \delta_\zeta \tag{6.34}$$

$$\omega^2 = \omega_o^2 + W_{\omega^2}\delta_{\omega^2} \tag{6.35}$$

Introduce these terms to the state equation for $\ddot{y}$ in (6.27).

$$\begin{aligned}
\ddot{y} &= -\omega^2\eta - 2\zeta\omega\dot{\eta} + u \\
&= -\left(\omega_o^2 + W_{\omega^2}\delta_{\omega^2}\right)\eta - 2\left(\zeta_o + W_\zeta\delta_\zeta\right)\omega_o\dot{\eta} + u \\
&= -\omega_o^2\eta - 2\zeta_o\omega_o\dot{\eta} + u - \delta_\zeta 2W_\zeta\omega_o\dot{\eta} - \delta_{\omega^2}W_{\omega^2}\eta \\
&= -\omega_o^2\eta - 2\zeta_o\omega_o\dot{\eta} + u + \delta_\zeta z_\zeta + \delta_{\omega^2}z_{\omega^2} \\
&= -\omega_o^2\eta - 2\zeta_o\omega_o\dot{\eta} + u + w_\zeta + w_{\omega^2}
\end{aligned} \tag{6.36}$$

This formulation uses only linear terms in the uncertainty parameters and so it can be considered for μ analysis. As usual, the explicit dependence on the uncertainty parameters is replaced by signals that relate the nominal dynamics to the uncertainty through feedback. These signals use the relationship $w_\zeta = \delta_\zeta z_\zeta$ and $w_{\omega^2} = \delta_{\omega^2}z_{\omega^2}$ where z_ζ and z_{ω^2} are fictitious outputs of the nominal plant.

$$z_\zeta = -2W_\zeta\omega_o\dot{\eta} \tag{6.37}$$

$$z_{\omega^2} = -W_{\omega^2}\eta \tag{6.38}$$

The resulting model can be realized as a state-space system.

$$\begin{bmatrix} \dot{\eta} \\ \ddot{\eta} \\ \hline z_\zeta \\ z_{\omega^2} \\ y \end{bmatrix} = \left[\begin{array}{cc|ccc} 0 & 1 & 0 & 0 & 0 \\ -\omega_o^2 & -2\zeta_o\omega_o & 1 & 1 & 1 \\ \hline -2\omega_o & 0 & 0 & 0 & 0 \\ -2\omega_o & -2\zeta_o & 0 & 0 & 0 \\ 1 & 0 & 0 & 0 & 0 \end{array}\right] \begin{bmatrix} \eta \\ \dot{\eta} \\ \hline w_\zeta \\ w_{\omega^2} \\ u \end{bmatrix} \tag{6.39}$$

Figure 6.5 demonstrates the feedback relationship between the state-space model of the plant P given in (6.39) and the uncertainty operators δ_ζ and δ_ω.

Parametric uncertainty associated with modal parameters can also be approximated by considering uncertainties in an eigenspace model. These models may be generated by performing a state transformation on the modal modal such that the state matrix contains real and imaginary elements of its eigenvalues. The eigenspace form of the model is often particularly useful because it easily shows information about stability that results from pole location; thus, it is logical to consider associated uncertainty with these types of models.

Consider an eigenspace formulation of the single-mode model given in (6.27) that is derived by transforming the states η and $\dot{\eta}$ into new states defined as x_1 and x_2.

$$\begin{bmatrix} \dot{x}_1 \\ \dot{x}_2 \\ \hline y \end{bmatrix} = \left[\begin{array}{cc|c} r & i & b_1 \\ -i & r & b_2 \\ \hline c_1 & c_2 & 0 \end{array}\right] \begin{bmatrix} x_1 \\ x_2 \\ \hline u \end{bmatrix} \tag{6.40}$$

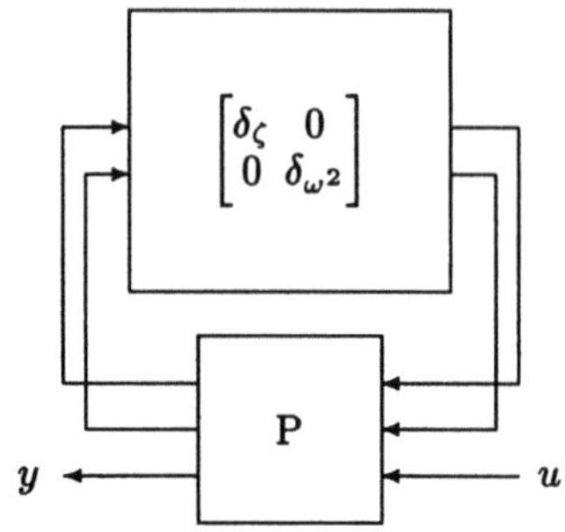

Fig. 6.5: Nominal Plant and Approximation to Parametric Uncertainty in Damping and Squared Natural Frequency

The elements r and i in the state matrix are chosen to represent the real and imaginary components of the poles of the system; however, they are directly related to the modal parameters because the natural frequency of the mode is $\omega = \sqrt{r^2 + i^2}$ and the damping is $\zeta = -r/\omega$. Thus, introducing uncertainty in the r and i elements allows consideration of uncertainty in the modal natural frequency and damping parameters.

Define uncertainty parameters $\delta_r \in \mathbf{R}$ and $\delta_i \in \mathbf{R}$ as multiplicative parametric uncertainties associated with nominal matrix parameters r_o and i_o. Additionally, define weights, W_r and W_i, for each uncertainty.

$$r = r_o(1 + W_r\delta_r) \tag{6.41}$$
$$i = i_o(1 + W_i\delta_i) \tag{6.42}$$

Extract the uncertainty parameters from the update equations for state x_1 by replacing the uncertain terms with external signals.

$$\begin{aligned}
\dot{x_1} &= rx_1 + ix_2 \\
&= r_o(1 + W_r\delta_r)x_1 + i_o(1 + W_i\delta_i)x_2 \\
&= r_o x_1 + i_o x_2 + r_o W_r \delta_r x_1 + i_o W_i \delta_i x_2 \\
&= r_o x_1 + i_o x_2 + \delta_r z_1 + \delta_i z_3 \\
&= r_o x_1 + i_o x_2 + w_1 + w_3
\end{aligned} \tag{6.43}$$

Perform a similar derivation on the state equation for x_2.

$$\begin{aligned} x_2 &= -ix_1 + rx_2 \\ &= -i_o(1+W_i\delta_i)x_1 + r_o(1+W_r\delta_r)x_2 \\ &= -i_ox_1 + r_ox_2 - i_oW_i\delta_ix_1 + r_oW_r\delta_rx_2 \\ &= -i_ox_1 + r_ox_2 - \delta_iz_4 + \delta_rz_2 \\ &= -i_ox_1 + r_ox_2 - w_4 + w_2 \end{aligned} \tag{6.44}$$

The additional signals $w_1, ..., w_4$ and $z_1, ..., z_4$ are introduced as feedback signals to relate the nominal dynamics and the uncertainty parameters. The signals $z_1, ..., z_4$ are output from the plant like measurements and they are functions of the states.

$$z_1 = W_rr_ox_1 \tag{6.45}$$

$$z_2 = W_rr_ox_2 \tag{6.46}$$

$$z_3 = W_ii_ox_2 \tag{6.47}$$

$$z_4 = W_ii_ox_1 \tag{6.48}$$

The vector of plant outputs, $z = \{z_1, z_2, z_3, z_4\}$, are related to the vector of plant inputs, $w = \{w_1, w_2, w_3, w_4\}$, through the relationship $w = \Delta_A z$ where the uncertainty operator Δ_A contains repeated blocks of the parametric uncertainty.

$$\Delta_A = \begin{bmatrix} I_{2\times2}\delta_r & 0 \\ 0 & I_{2\times2}\delta_i \end{bmatrix} \tag{6.49}$$

A state-space realization in eigenspace form can be written that contains these additional input and output signals to relate the structured uncertainty operator.

$$\begin{bmatrix} \dot{x}_1 \\ \dot{x}_2 \\ \hline z_1 \\ z_2 \\ z_3 \\ z_4 \end{bmatrix} = \left[\begin{array}{cc|ccccc} r_o & i_o & 1 & 0 & 1 & 0 & b_1 \\ -i_o & r_o & 0 & 1 & 0 & -1 & b_2 \\ \hline W_rr_o & 0 & 0 & 0 & 0 & 0 & 0 \\ 0 & W_rr_o & 0 & 0 & 0 & 0 & 0 \\ 0 & W_ii_o & 0 & 0 & 0 & 0 & 0 \\ W_ii_o & 0 & 0 & 0 & 0 & 0 & 0 \end{array}\right] \begin{bmatrix} x_1 \\ x_2 \\ \hline w_1 \\ w_2 \\ w_3 \\ w_4 \\ u \end{bmatrix} \tag{6.50}$$

The block diagram that describes the linear fractional transformation used for robust analysis of this system that relates the nominal plant P and the uncertainty parameters through feedback is given in Figure 6.6.

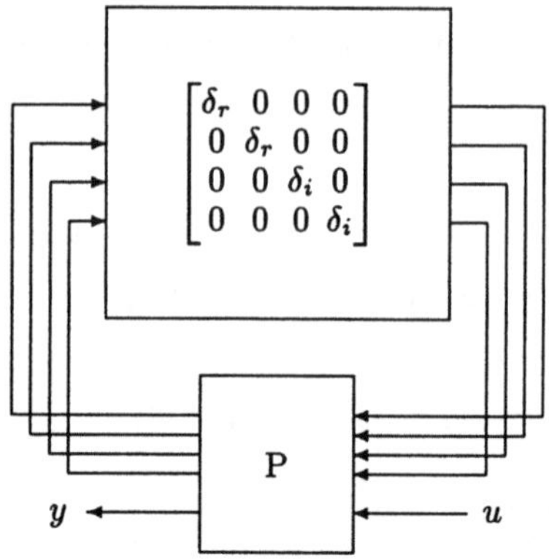

Fig. 6.6: Nominal Plant and Parametric Uncertainty in Eigen Elements

6.4 Dynamic Uncertainty

Dynamic uncertainty operators are often associated with aeroelastic models to account for modeling errors that are not efficiently described by parametric uncertainty. Unmodeled dynamics and inaccurate mode shapes are examples of modeling errors which can be described with less conservatism by dynamic uncertainties than with parametric uncertainties. These dynamic uncertainties are typically complex to represent errors in both magnitude and phase of signals.

Consider a system P having two modes with natural frequencies at 4 *rad/s* and 30 *rad/s*.

$$P = \left(\frac{16}{s^2 + .4s + 16}\right)\left(\frac{900}{s^2 + .6 + 900}\right) \tag{6.51}$$

Define a nominal model P_o which will be used for stability analysis of the system but does not model the high frequency mode of P.

$$P_o = \frac{16}{s^2 + .4s + 16} \tag{6.52}$$

The large difference in natural frequency between the high and low frequency modes of P precludes parametric uncertainty associated with the low frequency mode of P_o to be a reasonable description of the modeling errors in P_o. The magnitude of any parametric uncertainty associated with the low frequency mode would need to be extremely large to account for the unmodeled high frequency dynamics so the stability analysis would be relatively meaningless because the large uncertainty would imply the plant is not accurate at any frequencies.

A multiplicative uncertainty operator, $\Delta \in \mathbf{C}$, can be used to describe the high frequency modeling error without introducing the excessive conservatism resulting from a parametric uncertainty description. Associate a complex scalar weighting function $W(s) \in \mathbf{C}$ with this uncertainty to reflect the frequency-varying levels of modeling errors.

$$W = 100\frac{s + .5}{s + 500} \tag{6.53}$$

The set of plants $\mathcal{P}$ used for robust stability analysis is formulated to account for the range of dynamics as described by the norm bounded multiplicative uncertainty Δ.

$$\mathcal{P} = \{P_o(I + W\Delta) \ : \ \|\Delta\|_\infty \leq 1\} \tag{6.54}$$

The block diagram for robust stability analysis of $\mathcal{P}$ is given in Figure 6.7

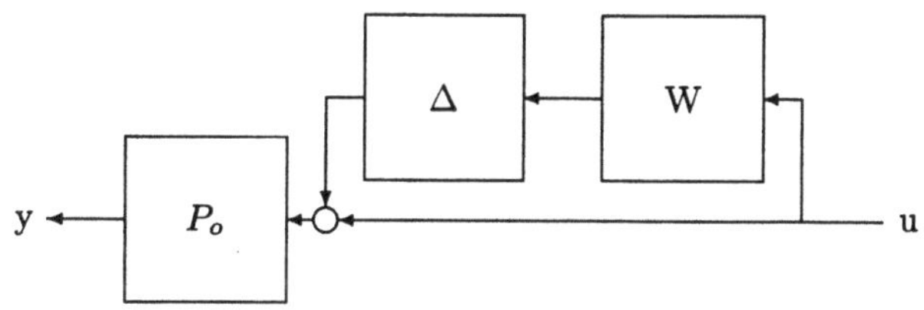

Fig. 6.7: Family of Plants $\mathcal{P} = P_o(I + \Delta W)$ with Input Multiplicative Uncertainty

Figure 6.8 presents the magnitude of the transfer function from input to output for P and P_o along with the maximum magnitude of $|P_o(I + W\Delta)|$ which is an upper bound for the output of P at each frequency. The multiplicative uncertainty is able to bound the modeling error at the high frequency mode without introducing excessive conservatism from large uncertainty associated with the low frequency mode.

Dynamic additive operators may also be required in the uncertainty description to account for errors that are not efficiently described by neither multiplicative nor parametric uncertainties. Modeling errors associated with a zero of the system dynamics is an example of an error which is best described by additive uncertainty. Multiplicative operators are not useful in this case since $P_o(\jmath\omega) = 0$ at the frequency ω associated with the zero of nominal model and correspondingly every member of the set of plants $P_o(\jmath\omega)(I + W\Delta) = 0$ at this frequency. Additive uncertainty allows the system output for some member of the set of plants to be nonzero even at frequencies of the zeros of the nominal plant.

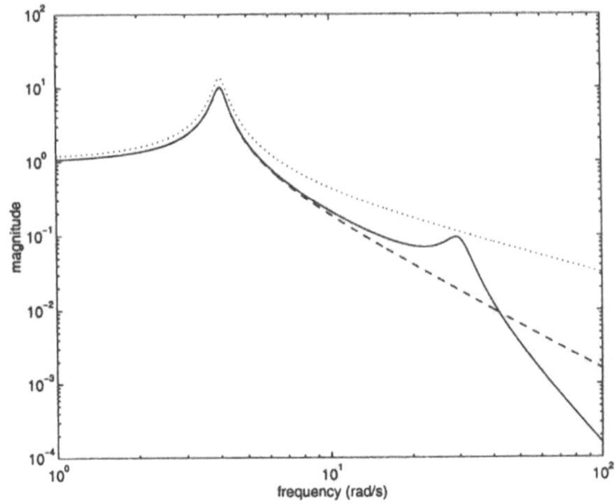

Fig. 6.8: Transfer Functions for Example System with Multiplicative Uncertainty : Plant P (—), Nominal Plant P_o (- - -), Maximum $|P_o(I + W\Delta)|$ at Each Frequency (...)

Consider a plant P with several poles and zeros.

$$P = \frac{16}{64}\left(\frac{s^2 + .48s + 64}{s^2 + .48s + 16}\right)\left(\frac{900}{s^2 + .6s + 900}\right) \tag{6.55}$$

Assume the nominal plant P_o of this plant is similar to P and has the correct number of poles and zeros but the coefficients of the system equations are incorrect.

$$P_o = \frac{16}{36}\left(\frac{s^2 + .36s + 36}{s^2 + .4s + 16}\right)\left(\frac{900}{s^2 + .6s + 900}\right) \tag{6.56}$$

Define an additive uncertainty operator $\Delta \in \mathbf{C}$ which is normalized by a complex scalar function W to reflect the frequency-varying levels of the modeling errors.

$$W = .1\frac{s + 50}{s + 5} \tag{6.57}$$

The set of plants $\mathcal{P}$ used for robust stability analysis is formulated to account for the range of dynamics as described by the norm bounded additive uncertainty Δ.

$$\mathcal{P} = \{P_o + W\Delta \ : \ ||\Delta||_\infty \leq 1\} \tag{6.58}$$

The block diagram for robust stability analysis of $\mathcal{P}$ is given in Figure 6.9

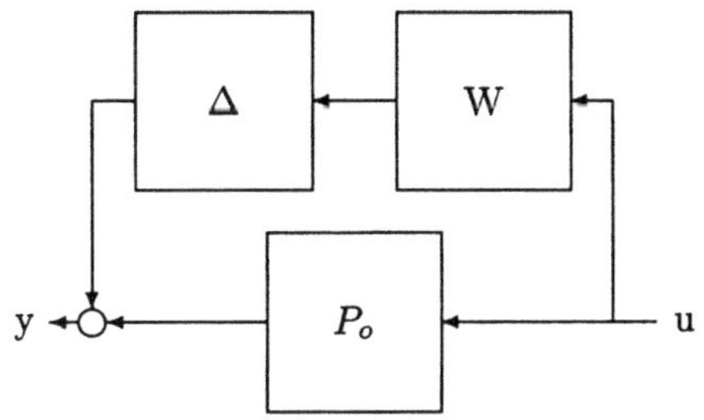

Fig. 6.9: Family of Plants $\mathcal{P} = P_o + W\Delta$ with Additive Uncertainty

Figure 6.10 presents the magnitude of the transfer function from input to output for P and P_o along with the maximum magnitude of $|P_o + W\Delta|$ at each frequency. The additive uncertainty bounds the modeling error at each frequency including the frequencies near the zero of the nominal plant since the output of P is bounded above by the maximum magnitude of the members of the set $\mathcal{P}$.

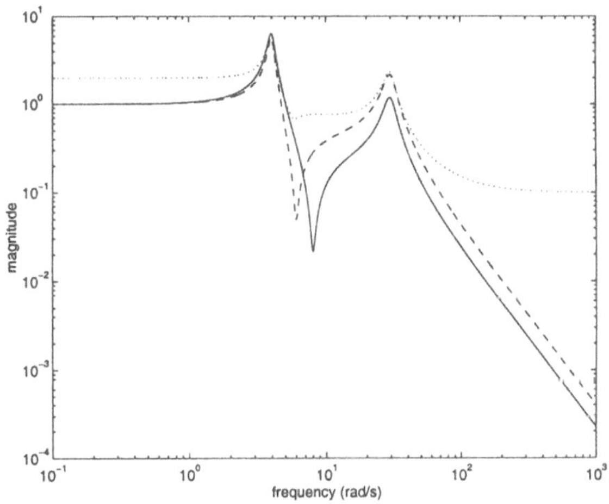

Fig. 6.10: Transfer Functions for Example System with Additive Uncertainty : Plant P (—), Nominal Plant P_o (- - -), Maximum $|P_o + W\Delta|$ at Each Frequency (...)

These multiplicative and additive uncertainties are particularly important when comparing an analytical transfer function with experimental transfer functions derived from flight data. Analytical models are often computed for a low range of frequencies since the high frequencies add complexity to the model but do not always affect the stability margins of the aircraft. The experimental data may indicate a high frequency mode that is not included in the analytical model so a frequency weighted dynamic multiplicative uncertainty can be associated with the model.

The issue of mode shape uncertainty is often encountered when comparing low frequency predicted dynamical responses with flight data since sensor measurements are directly affected by the mode shapes. Both multiplicative and additive uncertainties may be required to accurately model mode shape errors and account for inaccurate response levels, which may be higher than predicted at some frequencies but lower than predicted at others, along with inaccurate frequencies associated with poles and zeros.

6.5 Uncertainty Associated with Nonlinearities

The μ framework described in this paper utilizes linear operators to represent dynamical models and associated uncertainties but does not admit nonlinear operators. The μ framework is useful for analyzing stability of aircraft despite the constraints of linearity since physical systems, which are always nonlinear, can often be approximated by linear models to a high degree of accuracy. A classic example of this situation notes the linearized dynamics are often an acceptable representation of an aircraft operating near trim flight conditions so linear models work well in practice for control synthesis and stability analysis [1].

Nonlinear dynamics can not always be accurately described by a linearized dynamics model so the stability analysis should consider the effects of these nonlinearities [47]. The μ framework can associate linear uncertainty operators with linear models to describe the errors resulting from some types of unmodeled nonlinear dynamics. The uncertainty does not actually represent the nonlinearity; rather, it allows the linear system responses to vary with sufficient magnitude to bound a range of nonlinear system responses.

It is useful to separate the nonlinear dynamics which can not be linearized from the nonlinear dynamics which can be accurately represented by linear models. Separate uncertainty descriptions can be formulated for each dynamical block and the resulting operators combined using the LFT framework to formulate a single linear plant model with a structured uncertainty description. Figure 6.11 presents an example LFT system representation for a nominal plant P_o with associated uncertainty Δ and a nominal linear model N_o with associated uncertainty Δ_N representing a system element with nonlinear dynamics.

The system represented in Figure 6.11 is commonly used to describe the coupling between the aeroelastic dynamics and actuators affecting an aircraft through control surfaces. Actuators can display many types of nonlinear behaviors and should be considered in the aeroelastic stability analysis since

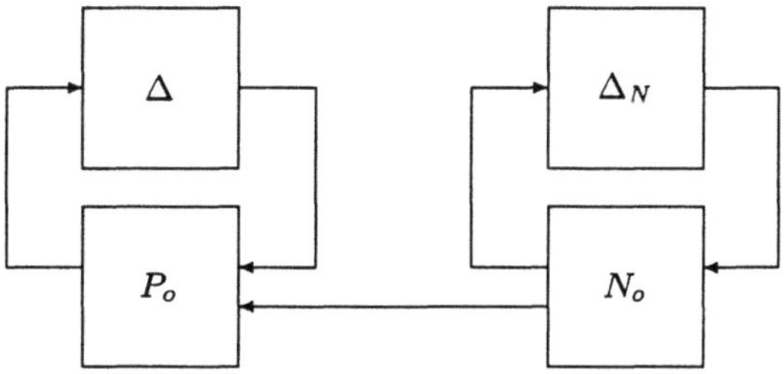

Fig. 6.11: Linear Fractional Transformation System with Nominal Models and Associated Uncertainty Operators

the control surfaces are continuously moving due to pilot and autopilot commands which maintain trim during flight [22]. The effects of errors in linear models resulting from unmodeled actuator dynamics such as nonlinear stiffness parameters or hysteresis functions can sometimes be described by a linear uncertainty operator.

Consider the response y of a nonlinear system N which models a system with a nonlinear stiffness corresponding to a hardening spring which is valid for the bounded input signal $u \in [-10, 10]$. Such a system may represent an element of an actuator model or a nonlinear structural model.

$$\begin{aligned} y &= N(u) \\ &= 2u + .02u^2 + .0082u^3 \end{aligned} \tag{6.59}$$

Define a linear nominal model N_o such that $y = N_o u$ approximates the response of N.

$$N_o = 2 \tag{6.60}$$

Associate an additive weighting operator Δ_N with N_o such that stability analysis considers the set of plants $\mathcal{N}$.

$$\mathcal{N} = \{N_o + \Delta_N \ : \ ||\Delta_N||_\infty \leq 1\} \tag{6.61}$$

Figure 6.12 demonstrates the maximum and minimum magnitudes of the responses of the set $\mathcal{N}$ are able to bound the response of the nonlinear system N for the range $u \in [-10, 10]$. These bounds are overly conservative throughout this operating range but they achieve the desired goal of describing errors in the linear system response resulting from the unmodeled nonlinearity.

A similar procedure can be used to describe error due to an unmodeled nonlinear softening spring. Consider the response y of a system represented by N which is valid for an input signal $u \in [-10, 10]$.

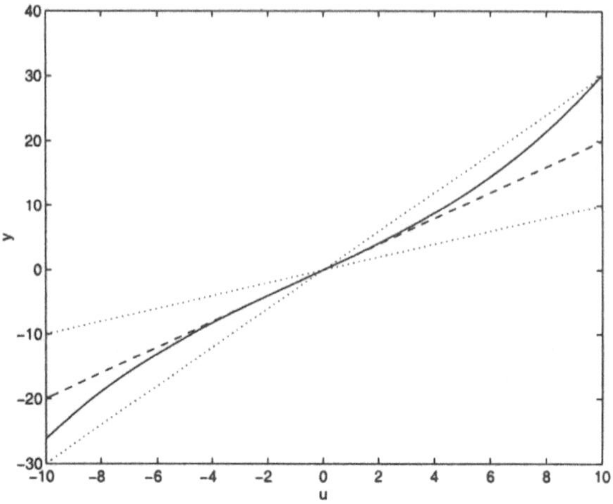

Fig. 6.12: System Responses for Hardening Spring Example : Nonlinear System N (—), Nominal Linear System N_o (- - -), Maximum and Minimum Bounds on Set $\mathcal{N} = \{N_o + \Delta_N\}$ (...)

$$
\begin{aligned}
y &= N(u) \\
&= 2u + .005u^2 - .0052u^3
\end{aligned}
\tag{6.62}
$$

Define a linear nominal model N_o such that $y = N_o u$ approximates the response of N.

$$
N_o = 2 \tag{6.63}
$$

Associate an additive weighting operator Δ_N with N_o such that stability analysis considers the set of plants $\mathcal{N}$.

$$
\mathcal{N} = \{N_o + \Delta_N \ : \ \|\Delta_N\|_\infty \leq 1\} \tag{6.64}
$$

Figure 6.13 demonstrates the maximum and minimum magnitudes of the responses of the set $\mathcal{N}$ are able to bound the response of the nonlinear system N for the range $u \in [-10, 10]$. These bounds are again overly conservative throughout this operating range but they achieve the desired goal of describing errors in the linear system response resulting from the unmodeled nonlinearity.

Another nonlinearity which commonly affects aeroelastic systems is hysteresis. The response of a hysteretic system may depend on the trend of the input signal such that an increasing input signal will generate a certain response but a decreasing input signal will generate a different response. Such

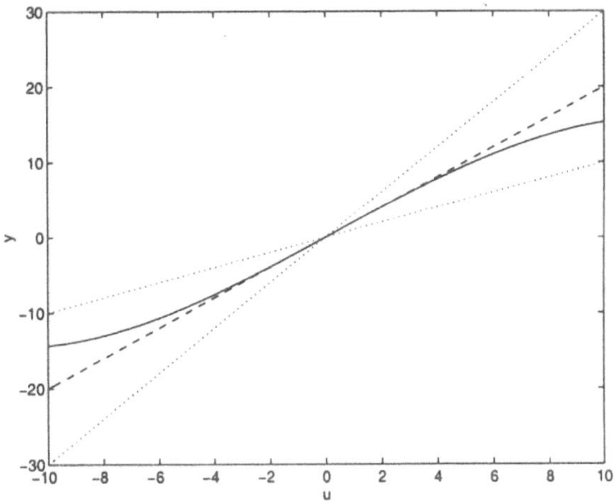

Fig. 6.13: System Responses for Softening Spring Example : Nonlinear System N (—), Nominal Linear System N_o (- - -), Maximum and Minimum Bounds on Set $\mathcal{N} = \{N_o + \Delta_N\}$ (...)

hysteresis dynamics are difficult to express as a simple mathematical formula so for illustrative purposes assume N is a nonlinear hysteresis function whose response y depends on the trend of the input signal.

Define a linear nominal model N_o whose response $y = N_o u$ approximates the response of the hysteretic N.

$$N_o = 2 \tag{6.65}$$

Associate an additive weighting operator Δ_N with N_o such that stability analysis considers the set of plants $\mathcal{N}$.

$$\mathcal{N} = \{N_o \mid \Delta_N \ : \ \|\Delta_N\|_\infty \leq 1\} \tag{0.00}$$

Figure 6.14 demonstrates the maximum and minimum magnitudes of the responses of the set $\mathcal{N}$ are able to bound the response of the nonlinear system N with the hysteresis for the range $u \in [-10, 10]$. Again, the bounds resulting from a linear uncertainty description are overly conservative throughout this operating range but they achieve the desired goal of describing errors due to the unmodeled hysteresis nonlinearity.

Explicitly constraining the operating region of the input signal $u \in [-10, 10]$ may be important to developing reasonable uncertainties to describe errors resulting from unmodeled nonlinearities. The errors in the linear model may grow excessively large when considering a large range of inputs so the uncertainty magnitude would need to grow excessively large also. The

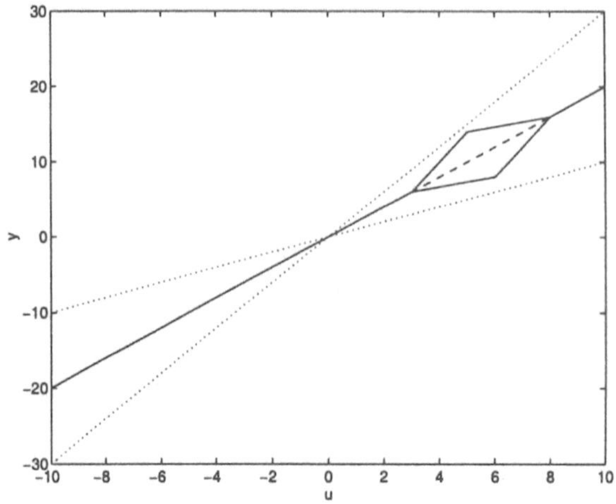

Fig. 6.14: System Responses for Hysteresis Example : Nonlinear System N (—), Nominal Linear System N_o (- - -), Maximum and Minimum Bounds on Set $\mathcal{N} = \{N_o + \Delta_N\}$ (...)

conservatism resulting from such a large uncertainty description may be unacceptable and require the input range to be constrained to more reasonable limits.

The uncertainty description, even for a constrained operating region, will usually be overly conservative when describing modeling errors for certain parts of the operating region. The uncertainty is able to bound the errors in Figures 6.12,6.13, and 6.14 but the maximum and minimum bound are clearly not optimal. Some amount of conservatism is expected when describing errors resulting from unmodeled nonlinearities since a linear model, whether a single plant or a set of plants, will usually not be an accurate representation of a nonlinear system which can not be linearized.

Also, there are several types of unmodeled nonlinearities that are frequently encountered in aircraft systems but are not easily described by linear uncertainty operators associated with the linear models. Examples of these types of nonlinearities include freeplay, deadband responses, friction, and rate limiting of actuators.

The robust stability margins computed from μ with respect to the linear uncertainty operators describing unmodeled nonlinear dynamics will always be somewhat suspect since this approach can not consider stability properties unique to nonlinear systems such as bifurcation points and limit cycle behaviors [130, 137]. This limits the usefulness of the μ method to systems for which the nonlinearities have small effects on the response and do not introduce nonlinear instabilities to the critical mechanism of a flutter or aeroservoelastic instability.

6.6 Uncertainty Associated with Flight Data

A theoretical model with an associated uncertainty description may be an accurate representation of the aeroelastic dynamics of an aircraft but responses from that model may not identically match flight data. Additional uncertainties can be associated with the model to describe errors which are observed between the predicted responses and the measured responses from a commanded excitation to the aircraft. These uncertainties do not necessarily indicate errors in the model; rather, they indicate errors in the process used to generate aeroelastic responses and measure flight data.

One source of error is an incorrect assumption of excitation force used to generate the predicted and measured responses. The measured excitation force associated with the flight data may not correctly account for poor hardware performance and non-uniform spectral distribution of the force. Also, inexactly phased excitation between multiple force mechanisms may excite modes which are not anticipated by a theoretical analysis. A frequency-varying dynamic uncertainty can be associated with the force input of the analytical model to describe errors in the excitation.

The phenomenon of nonrepeatibility can cause discrepancies between predicted and measured responses from multiple occurances of excitation signals. Nonrepeatibility affects flight data by introducing slight variations in responses even for data recorded at identical flight conditions with identical excitation signals. This unexplained behavior may result from some unmodeled nonlinear dynamic or inexact excitation which is not correctly measured. External disturbances such as wind gusts or turbulence may introduce an unmodeled dynamic that inconsistently affects the aircraft responses. A frequency-varying dynamic uncertainty may be associated with the model to describe nonrepeatible data variations.

Another source of error between predicted and measured responses is an incorrect assumption of flight condition. Flight data sets are sometimes generated at test points which attempt to maintain a constant flight condition to match the data sets predicted from a model describing the aeroelastic dynamics at that same flight condition. Slight variations in flight conditions while the experimental response is measured may cause some discrepancy between the predicted response and the flight data. Parametric uncertainty associated with the unsteady aerodynamic model can be used to account for these errors since flight condition variations only affect the aerodynamic model and not the structural model.

The model may accurately represent the mode shapes of the aircraft but have a poor representation of the sensor locations. The responses measured by sensors are inherently dependent on their location with respect to mode shapes to determine the magnitude and phase of the signal. Additional errors in magnitude and phase are introduced when considering transfer function estimates generated by signals which violate assumptions used in computational algorithms. A frequency-varying dynamic uncertainty can be associated with the output of the plant model to describe errors in both mode shape and sensor location predictions.

The choice of signal processing algorithms can also introduce errors between predicted and measured responses. The Fourier transform, which is the traditional tool for signal processing, assumes several characteristics of the data which may be violated with transient response aeroelastic data [26]. Filtering and wavelet based algorithms may be used to reduce errors introduced by signal processing and reduce conservatism in the resulting stability margins [23, 44]. Parameter uncertainty associated with the modal parameters of the linear model may be used to describe some errors in the natural frequencies and dampings observed with flight data analyzed by incorrect algorithms. Dynamic uncertainty may also be required to describe errors introduced by leakage and aliasing effects.

CHAPTER 7

INCORPORATING FLIGHT DATA

7.1 Model Validation

Robust stability analysis considers the stability of a system subject to a set of modeling errors and perturbations represented by a norm bounded set $\mathbf{\Delta}$. A logical question that arises is how to reasonably determine this set. This issue is important for computing robust aeroservoelastic stability margins since μ may be overly conservative if the uncertainty is excessive beyond the true modeling errors and may be overly optimistic if the uncertainty does not sufficiently account for the true modeling errors.

Model validation algorithms may be used to indicate if an uncertainty description is reasonable with respect to flight data [85]. These algorithms consider if a set of data measurements could have been generated by a proposed model which includes the nominal dynamics and associated uncertainty operators along with noise and disturbance signals.

Uncertainty operators associated with nominal plant models specifically in the linear fractional transformation framework can be considered by validation algorithms [28]. Frequency-domain algorithms are generated which consider the model validation problem in the context of control design [122]. Time-domain approaches are also developed which may be solved with convex optimization algorithms for certain uncertainty structures [105].

A model validation procedure has been formulated that uses a μ condition to determine if an uncertainty model is invalidated [70]. This procedure uses frequency-domain transfer function data to determine if some perturbation $\Delta \in \mathbf{\Delta}$ to the nominal plant could enable the theoretical model to produce measurements that match the experimental data.

The model validation question as applied to a model, P, with an associated uncertainty description, Δ, that is related through a linear fractional transformation is given in Question 7.1.1.

Question 7.1.1 *(uncertainty validation) Is there some frequency-varying $\Delta \in \mathbf{\Delta}$ for Figure 7.1 such that $F_u(P, \Delta)$ could generate the set of observed data y and u?*

The issue of model validation is inherently different than that of system identification. Consider that the validation question does not attempt to find the required Δ; rather, it simply determines if a valid $\Delta \in \mathbf{\Delta}$ exists. This is an important consideration when attempting to generate aeroelastic models by analyzing noisy flight data.

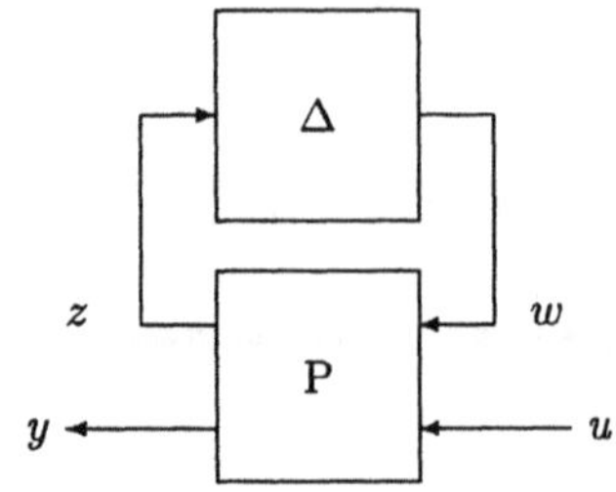

Fig. 7.1: Linear Fractional Transformation System for Robust Stability Analysis and Model Validation with Forcing and Measurement Signals

Consider the block diagram for robust stability analysis of systems with measurement y and forcing u signals in Figure 7.1.

Define $P(s) \in \mathbf{C}^{n_o+n_z \times n_i+n_w}$ as a stable transfer function system matrix such that $P \in \mathcal{RH}_\infty$. Partition this matrix into four elements such that $P_{11}(s) \in \mathbf{C}^{n_z \times n_w}$, $P_{22}(s) \in \mathbf{C}^{n_o \times n_i}$ and $P_{12}(s)$ and $P_{21}(s)$ are of appropriate size. Note with this formulation that $P_{22}(s)$ is the nominal plant transfer function in the absence of uncertainty.

$$P = \begin{bmatrix} P_{11} & P_{12} \\ P_{21} & P_{22} \end{bmatrix} \tag{7.1}$$

Robust stability analysis of this system is determined using the transfer function P_{11} which contains the feedback relationship between the plant dynamics and the uncertainty operator. For example, the robust stability condition given by Theorem 3.3.3 as applied to the system in Figure 7.1 requires that $\mu(P_{11}) < 1$ for P to be robustly stable to the set of uncertainties represented by all $\Delta \in \mathbf{\Delta}$.

The model validation condition utilizes these elements of P along with the finite0energy frequency-domain signals $y, u \in \mathcal{L}_2$ where the measurements are complex-valued scalar functions $y, u \in \mathbf{C}$. Formulate the following two matrices.

$$\hat{P}_{12} = P_{12}u \tag{7.2}$$
$$\hat{P}_{22} = P_{22}u - y \tag{7.3}$$

The following theorem formulates the model validation test using μ [70].

Theorem 7.1.2 *Given measurements y generated by inputs u, then the system P with associated uncertainty set $\mathbf{\Delta}$ is not invalidated if*

$$\mu\left(P_{11} - \hat{P}_{12}\hat{P}_{22}^{-1}P_{21}\right) > 1 \tag{7.4}$$

This condition may seem counter intuitive in that the desired condition for validation is μ greater than one while the robust stability condition seeks a value less than one. This may be explained by considering the following relationship of $y = F_u(P, \Delta)u$ as shown in Figure 7.1.

$$\begin{aligned} 0 &= [P_{22}u - y] + P_{21}\Delta\left(I - P_{11}\Delta\right)^{-1}[P_{12}u] \\ &= \hat{P}_{22} + P_{21}\Delta\left(I - P_{11}\Delta\right)^{-1}\hat{P}_{12} \end{aligned} \tag{7.5}$$

Define the plant $\hat{P} = \left\{P_{11}, \hat{P}_{12}, P_{21}, \hat{P}_{22}\right\}$. A value of $\mu(\hat{P}) < 1$ implies this system is robustly stable to all perturbations $\Delta \in \mathbf{\Delta}$. This robust stability of $\mu(\hat{P}) < 1$ also implies that the loop gain $F_u(\hat{P}, \Delta)$ is not singular for any value $\Delta \in \mathbf{\Delta}$. Satisfying $\mu(\hat{P}) < 1$ would thus contradict the relationship presented in the above equations which requires $F_u(\hat{P}, \Delta)$ to be singular to satisfy the input to output relationship of the data. In this respect, the model validation test is actually an inverted robust stability test.

It is important to note the phrase "model validation" may actually be misleading. No analytical model can ever be truly validated by considering a finite set of experimental data. A model may not be invalidated by the data but there is no guarantee a different data set could not be generated from the physical system that invalidates the model. The finite sets of measurement data can not record the response of the system to an infinite number of input signals subject to an infinite number of initial conditions. Theorem 7.1.2 reflects this fact by explicitly stating the condition only determines if the model is invalidated by the data. It is assumed the system is nearly linear and the data sufficiently represents the behavior of the dynamics so Theorem 7.1.2 represents a model validation test.

7.2 Determining Levels of Uncertainty

The value of μ computed using Theorem 7.1.2 can be interpreted as a measure of how reasonable is the uncertainty description. A value of $\mu = 2$ implies the uncertainty may be reduced by a factor of 2 before the model is invalidated. Alternatively, a value of $\mu = .5$ implies twice as much uncertainty is required for the model not to be invalidated. This interpretation of μ as relating the size of allowable perturbations emphasizes the relationship of the model validation condition to a robust stability condition.

This concept of model validation allows flight data to be used to generate reasonable norm bounds for an uncertainty description. The following algorithm can be used to determine a sufficient level of uncertainty required such that the model is not invalidated by multiple data sets. A small scalar $\alpha > 1$ is chosen to scale the uncertainty set and increase the amount of allowable errors if the size of $\mathbf{\Delta}$ is not sufficient.

Algorithm 7.2.1 *(Model Validation)*

Given • *frequency-domain data* $\{y_1, y_2, \ldots, y_n\}$ *and* $\{u_1, u_2, \ldots, u_n\}$
• *frequency-domain model* $P = \{P_{11}, P_{12}, P_{21}, P_{22}\}$
• *uncertainty set setU*

Define • *scalar* $\alpha > 1$ *for level updates*

Then

```
valid = FALSE
while (valid == FALSE) {
        valid=TRUE
        for i = 1 : n {
                P̂12 = P12 ui
                P̂22 = P22 ui − yi
                if μ(P11 − P̂12 P̂22^{-1} P21) < 1 {
                        valid = FALSE
                }
        }
        P11 = αP11          ( equivalent to Δ = αΔ)
}
Δ is the required norm bounded uncertainty
```

$$\hat{P}_{12} = P_{12}u_i, \quad \hat{P}_{22} = P_{22}u_i - y_i, \quad \mu\left(P_{11} - \hat{P}_{12}\hat{P}_{22}^{-1}P_{21}\right) < 1$$

The norm bound on the uncertainty set $\boldsymbol{\Delta}$ is not actually increased with this method as denoted by the parenthesis around the comment statement $\boldsymbol{\Delta} = \alpha\boldsymbol{\Delta}$. It would be difficult to scale the uncertainty and retain these scalings throughout the procedure; instead, the algorithm is simplified by always considering the uncertainty is scaled such that $\mu < 1$ is the desired result. This scaling of the uncertainty is actually accomplished by the statement $P_{11} = \alpha P_{11}$ which scales the feedback signals between the plant and uncertainty operators. The norm bound for the uncertainty needed to ensure the uncertainty levels are not invalidated by the data is scaled into the plant by the parameter α. Robust flutter margins can be computed directly from the new scaled plant using Algorithm 8.1.6 with a desired $\mu < 1$ condition.

Algorithm 7.2.1 is a straightforward implementation of the model validation test that only compares the μ values with one to determine if the model is invalidated. More sophisticated algorithms could utilize the value of μ to determine the factor α to scale the uncertainty. Also, the values of μ across frequency could be exploited because frequencies with low μ value indicate areas where the uncertainty set is least conservative while frequencies with high μ are overly conservative. The scaling α could vary with frequency to reflect this information.

The situation may arise when the initial value chosen for the norm bound of the uncertainty set may be overly conservative. In this situation the model validation condition will pass for each data set during the first processing of the outer loop. The initial norm bound can simply be decreased by some level consistent with the lowest μ value computed during the validation checks and Algorithm 7.2.1 can be run again to compute a less conservative uncertainty description which does not invalidate the model.

Theorem 7.1.2 is only valid for scalar data signals generated by systems with a single input and single output. Multiple data signals may be considered by applying Theorem 7.1.2 to each combination of single input and output signals. Algorithm 7.2.1 may still be used by simply including outer loops to cycle over the number of input and measurement signals.

The model validation algorithm using Theorem 7.1.2 is a departure from traditional methods of analyzing flight data to assess accuracy of an analytical model. The most widely used algorithms for analyzing flight data estimate natural frequencies and modal damping. The model validation procedure using μ considers the response dynamics at each frequency without explicitly comparing modal properties. This procedure enhances the ability of the μ method to analyze flutter stability without requiring damping estimates.

Another interpretation of Algorithm 7.2.1 uses the uncertainty $\mathbf{\Delta}$ to bound the magnitude and phase of the possible transfer functions generated by the family of plants $F_u(P, \mathbf{\Delta})$. μ ensures the experimental data transfer function lies within these analytical bounds at every frequency.

This interpretation is easily shown using a simple example. Consider a system that is described as a single-mode model with mass mass $m = 12$, damping $c = 1$, and stiffness $k = 270$. The plant P_{true} is used to represent the displacement y computed in response to a disturbance u through the relationship $y = P_{true}u$.

$$P_{true} = \left[\begin{array}{cc|c} 0 & 1 & 0 \\ -22.5 & -0.8 & 1 \\ \hline 1 & 0 & 0 \end{array}\right] \tag{7.6}$$

Assume there is a model of this system but the actual value of stiffness is not known; instead, a nominal value of $k_o = 200$ represents the best estimate of the true stiffness. The equation of motion must account for this potential modeling error by including uncertainty.

$$\begin{aligned} 0 &= m\ddot{x} + c\dot{x} + kx + u \\ &= m\ddot{x} + c\dot{x} + (k_o + W_k\delta_k)x + u \end{aligned} \tag{7.7}$$

A linear fractional transformation can be formulated for this system that relates a state-space realization of the nominal dynamics and the uncertainty perturbation through signals z and w. Define P as this nominal state-space plant whose outputs are the uncertainty output z and the plunge displacement y whereas the inputs are the uncertainty feedback w and input u.

$$P = \left[\begin{array}{cc|cc} 0 & 1 & 0 & 0 \\ -16.7 & -0.8 & 1 & 1 \\ \hline 0.08W_k & 0 & 0 & 0 \\ 1 & 0 & 0 & 0 \end{array}\right] \tag{7.8}$$

The transfer function from input to displacement is shown in Figure 7.2 for the P_{true} and for P with the nominal value of stiffness.

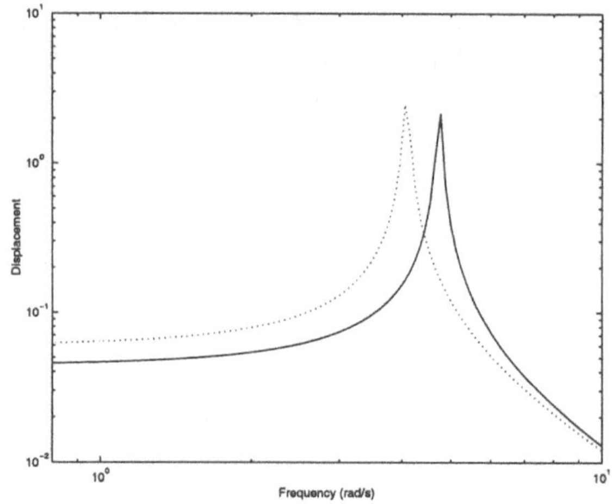

Fig. 7.2: Transfer Function for P_{true} (—) and nominal P (– – –)

Figure 7.2 demonstrates that the nominal plant model does not adequately describe the true dynamics; however, the amount of uncertainty that is needed to represent the modeling errors is not immediately obvious. Simply including a large amount of uncertainty will ensure the robust model includes the true model but also probably lead to an overly conservative stability margin; instead, apply Algorithm 7.2.1 to determine a value of W_k that introduces a reasonable level of uncertainty.

Consider the range of transfer functions that result by choosing a initial level of uncertainty such that $W_k = 33$. The maximum value of the transfer function at every frequency that searches over all possible transfer functions is compared with the transfer function of P_{true} in Figure 7.3.

Figure 7.3 easily demonstrates why the model validation approach is conveniènt for determining levels of model uncertainty. The feature to note is that the transfer function from the true plant lies outside the maximum value of the transfer functions admitted by the initial uncertainty weighting of $W_k = 33$ at the natural frequency of the mode near 47 *rad/sec*. Thus, the uncertainty level needs to be increased.

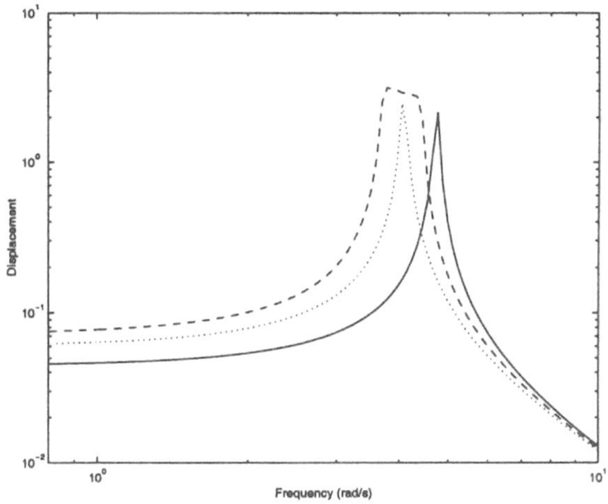

Fig. 7.3: Transfer Function for P_{true} (—), nominal P (– – –), and maximum value of $F_u(P, \delta_k)$ for $W_k = 33$ (...)

Consider the similar plot that results by increasing the weighting associated with uncertainty to the value of $W_k = 100$ as shown in Figure 7.4. This plot shows that the new level of uncertainty is large enough that the range of transfer functions for the nominal plants is sufficient to bound the transfer function data from the true plant.

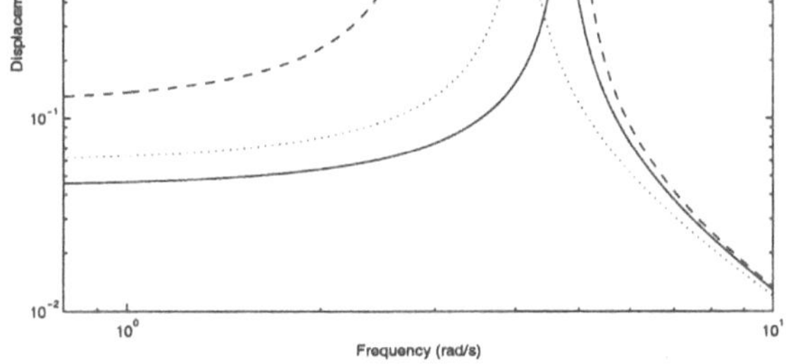

Fig. 7.4: Transfer Function for P_{true} (—), nominal P (– – –), and maximum value of $F_u(P, \delta_k)$ for $W_k = 100$ (...)

CHAPTER 8

STABILITY MARGINS

8.1 Robust Aeroelastic Stability Margins

Aeroelastic stability margins, or flutter margins, can be computed using μ to answer the nominal flutter problem posed in Question 5.2.1 and the robust flutter problem posed in Question 5.3.1. The value of μ is a sufficient direct indication of the flutter margin for the nominal system while additional information about the norm bound on the uncertainty is required to derive the robust flutter margin.

The nominal aeroelastic model is formulated for stability analysis in the μ framework in Section 5.2 by introducing a perturbation $\delta_{\overline{q}}$ to dynamic pressure. A nominal flutter margin is computed to answer Question 5.2.1 by considering the smallest value of this perturbation which destabilizes the model. The nominal flutter pressure can be directly calculated by computing μ which respect to the perturbation operator $\delta_{\overline{q}}$.

The exact value of μ can be analytically formulated since a closed form solution for μ with respect to a single real scalar operator is known. This solution is the maximum spectral radius of the frequency-varying transfer function matrix; however, the spectral radius is a discontinuous function of frequency so computational algorithms that search over a finite set of frequency points may not guarantee the true value of μ is is computed.

A small amount of complex uncertainty may be added to the real uncertainty which allows a continuous μ function to be analyzed but introduces unrealistic conservatism. Alternatively, a heuristic robustness indicator can be substituted for μ which considers stability over a frequency segment [41, 42].

A relatively simple approach may be used to compute μ with respect to a single real parameter by considering the destabilizing value of the parameter. A search over the parameter space will result in computation of μ and compute the desired flutter margin. Lemma 8.1.1 presents the principle behind this approach.

Lemma 8.1.1 *Given the plant P derived at nominal dynamic pressure $\overline{q}_o$ with a perturbation to dynamic pressure $\delta_{\overline{q}}$ as in Figure 5.2, define δ.*

$$\delta = \min_{\delta_{\overline{q}}>0} \{\delta_{\overline{q}} \ : \ F_u(P, \delta_{\overline{q}}) \textit{ is unstable}\}$$

Then $\mu(P) = \frac{1}{\delta}$ so $\overline{q}_{nom} = \overline{q}_o + \delta$ is the nominal flutter pressure and

$$\Gamma_{nom} = \delta$$

represents the nominal flutter margin answering Question 5.2.1.

Proof :
This result is immediately obvious using Definition 3.3.1 for μ with respect to a scalar uncertainty parameter $\delta_{\overline{q}}$. μ is the inverse of the smallest destabilizing perturbation and δ is computed as the smallest value of $\delta_{\overline{q}}$ which destabilizes P so $\mu(P) = \frac{1}{\delta}$.
□

Lemma 8.1.1 indicates a computational strategy to compute a nominal flutter margin that does not require a search over a set of frequency points. The flutter pressure is found by increasing values of $\delta_{\overline{q}}$ until an eigenvalue of the state matrix of $F_u(P, \delta_{\overline{q}})$ has a negative real part indicating $F_u(P, \delta_{\overline{q}})$ is unstable. Algorithm 8.1.2 demonstrates a bisection search implementation which efficiently computes upper and lower bounds on the minimum destabilizing $\delta_{\overline{q}}$ perturbation to within a desired level of accuracy.

Algorithm 8.1.2 *(nominal flutter margin)*

Given • *plant P at dynamic pressure $\overline{q}_o$*
• *operator $\delta_{\overline{q}}$ related to P as in Figure 5.1*

Define • *scalars $\delta_{upper} > \delta_{lower} > 0$ to bound $\delta_{\overline{q}}$*
• *scalar $\epsilon > 0$ for accuracy*

Then

while $(\delta_{upper} - \delta_{lower} > \epsilon)$ {
  $\delta_{\overline{q}} = \frac{1}{2}(\delta_{upper} + \delta_{lower})$
  if $(F_u(P, \delta_{\overline{q}})$ *has an unstable pole) then* $\delta_{upper} = \delta_{\overline{q}}$
  else $\delta_{lower} = \delta_{\overline{q}}$
}
$\overline{q}_{nom} = \overline{q}_o + \delta_{upper}$
$\Gamma_{nom} = \delta_{upper}$

Robust flutter margins which address Question 5.3.1 can not be computed using Algorithm 8.1.2 since an additional search over the set of uncertainty operators $\boldsymbol{\Delta}$ must be included. These margins must be computed using the augmented plant P which includes feedback signals relating the perturbation to dynamic pressure and the uncertainty description as in Figure 5.2. Define the block structured set of operators $\overline{\boldsymbol{\Delta}}_{\delta_{\overline{q}}}$ which considers a particular perturbation $\delta_{\overline{q}}$ and set of operators $\boldsymbol{\Delta}$ describing model uncertainty.

$$\overline{\boldsymbol{\Delta}}_{\delta_{\overline{q}}} = \left\{ \overline{\Delta}_{\delta_{\overline{q}}} \ : \ \overline{\Delta}_{\delta_{\overline{q}}} = \begin{bmatrix} \delta_{\overline{q}} & 0 \\ 0 & \Delta \end{bmatrix}, \Delta \in \boldsymbol{\Delta}, \|\Delta\|_\infty \leq 1 \right\} \tag{8.1}$$

There is a set of plant models $F_u(P, \overline{\mathbf{\Delta}}_{\delta_{\overline{q}}})$ for each value of $\delta_{\overline{q}}$ which defines the nominal plant at dynamic pressure $\overline{q} = \overline{q}_o + \delta_{\overline{q}}$ and variations to that plant caused by the set of uncertainties $\mathbf{\Delta}$. The robust flutter margin corresponds to the smallest perturbation $\delta_{\overline{q}}$ for which there exists an unstable plant $F_u(P, \overline{\Delta}_{\delta_{\overline{q}}})$ with $\overline{\Delta}_{\delta_{\overline{q}}} \in \overline{\mathbf{\Delta}}_{\delta_{\overline{q}}}$. If every member of the set of plants $F_u(P, \overline{\mathbf{\Delta}}_{\delta_{\overline{q}}})$ then $\overline{q} = \overline{q}_o + \delta_{\overline{q}}$ is not a flutter pressure and P formulated at $\overline{q}$ is robustly stable to the uncertainty description $\mathbf{\Delta}$.

The smallest destabilizing perturbation $\delta_{\overline{q}}$ corresponding to a robust flutter margin can be computed by a μ computation. The μ framework analyzes robustness with respect to a single structured operator so the operator set used to compute a robust flutter margin must contain the set of uncertainty operators $\mathbf{\Delta}$ along with a range of dynamic pressure perturbations. μ will compute the robustness of P with respect this operator set to find the smallest destabilizing perturbation to dynamic pressure and the smallest destabilizing uncertainty operator. Define the set $\overline{\mathbf{\Delta}}$ which contains $\overline{\mathbf{\Delta}}_{\delta_{\overline{q}}}$ sets for a norm bounded set of $\delta_{\overline{q}}$ operators.

$$\overline{\mathbf{\Delta}} = \left\{ \overline{\Delta} \; : \; \overline{\Delta} = \begin{bmatrix} \delta_{\overline{q}} & 0 \\ 0 & \Delta \end{bmatrix}, \; \Delta \in \mathbf{\Delta}, \; \|\Delta\|_\infty \leq 1, \; \|\delta_{\overline{q}}\|_\infty \leq 1 \right\} \tag{8.2}$$

Imposing the norm bound for $\delta_{\overline{q}}$ operators as $\|\delta_{\overline{q}}\|_\infty \leq 1$ may seem overly restrictive since the units of $\delta_{\overline{q}}$ is the same as the units of $\overline{q}$ in the model. This implies the set $\overline{\mathbf{\Delta}}$ considers the range of flight conditions $\overline{q} = \overline{q}_o \pm 1\ lb/ft^2$ for plants formulated by dynamic pressure in units of lb/ft^2. Such a small range of flight conditions is not useful for stability analysis unless $\overline{q}_o$ is extremely close to the flutter pressure. This limitation is avoided by introducing a weighting function $W_{\overline{q}}$ to the computation of $\overline{q}$.

$$\overline{q} = \overline{q}_o + W_{\overline{q}} \delta_{\overline{q}} \tag{8.3}$$

A weighting $W_{\overline{q}} > 1$ allows a large range of flight conditions to be considered despite the unity norm bound constraint on $\delta_{\overline{q}}$. This weighting is incorporated into the stability analysis by scaling the feedbacks signals between the $\delta_{\overline{q}}$ operator and the plant P to form the scaled plant $\overline{P}$.

$$\overline{P} = P \begin{bmatrix} W_{\overline{q}} & 0 \\ 0 & I \end{bmatrix} \tag{8.4}$$

A robust flutter margin is computed by analyzing $\mu(\overline{P})$ with respect to the set $\overline{\mathbf{\Delta}}$. The robust flutter pressure is determined by iterating over scalings $W_{\overline{q}}$ until the smallest pressure $\overline{q} = \overline{q}_o + W_{\overline{q}}$ is found for which the $\overline{P}$ is robustly stable to the set of uncertainties $\mathbf{\Delta}$. Theorem 8.1.3 formally demonstrates this concept.

Theorem 8.1.3 *Given the plant P derived at nominal dynamic pressure $\overline{q}_o$ with a perturbation to dynamic pressure $\delta_{\overline{q}}$ and set of uncertainty operators $\mathbf{\Delta}$ norm bounded by one arranged in the feedback relationship of Figure 5.2, define the plant $\overline{P}$ with real diagonal matrix $W_{\overline{q}}$ scaling the feedback signals relating $\delta_{\overline{q}}$ and P.*

$$\overline{P} = P \begin{bmatrix} W_{\overline{q}} & 0 \\ 0 & I \end{bmatrix}$$

Then $\overline{q}_{rob} = \overline{q}_o + W_{\overline{q}}$ is the robust flutter pressure and

$$\Gamma_{rob} = W_{\overline{q}}$$

represents the least conservative robust flutter margin that answers Question 5.3.1 if and only if $\mu(\overline{P}) = 1$.

Proof

$\Rightarrow$ (necessary)

The condition $\mu(\overline{P}) = 1$ implies the smallest destabilizing perturbation to $\overline{P}$ is described by some $\overline{\Delta} \in \overline{\mathbf{\Delta}}$ with $||\overline{\Delta}||_\infty = 1$ so there is no destabilizing $\Delta \in \mathbf{\Delta}$ and the smallest positive destabilizing perturbation to dynamic pressure is at least $\delta_{\overline{q}} = 1$ which corresponds to dynamic pressure $\overline{q} = \overline{q}_o + W_{\overline{q}}\delta_{\overline{q}} = \overline{q}_o + W_{\overline{q}}$. Thus, $\overline{P}$ is guaranteed to be robustly stable to the uncertainty set $\mathbf{\Delta}$ for any perturbation to dynamic pressure less than $W_{\overline{q}}$ so $W_{\overline{q}}$ is a robust flutter margin.

$\Leftarrow$ (sufficient)

Assume $\mu(\overline{P}) > 1$. Define real scalar $\alpha < 1$ such that $\mu(\overline{P}) = \frac{1}{\alpha}$ which implies, from Theorem 3.3.2, that $\overline{P}$ is robustly stable to all uncertainties $\overline{\Delta} \in \overline{\Delta}$ with $||\overline{\Delta}||_\infty < \alpha < 1$. Thus, $\overline{P}$ is not guaranteed to be stable over the entire range of modeling uncertainty defined by the unity norm bounded set $\overline{\mathbf{\Delta}}$ so this is not a valid robust flutter margin and does not answer Question 5.3.1.

Assume $\mu(\overline{P}) < 1$. Define real scalar $\alpha > 1$ such that $\mu(\overline{P}) = \frac{1}{\alpha}$ which implies, from Theorem 3.3.2, that $\overline{P}$ is robustly stable to all uncertainties $\overline{\Delta} \in \overline{\Delta}$ with $||\overline{\Delta}||_\infty < \alpha$. Thus, $\overline{P}$ is robustly stable to an uncertainty description larger than that defined by the unity norm bounded set $\overline{\mathbf{\Delta}}$ so this condition defines a valid flutter margin but it is not the least conservative robust flutter margin and does not answer Question 5.3.1.

□

The only difference between models $\overline{P}$ and P results from the scaling $W_{\overline{q}}$ which scales the feedback signals between $\overline{P}$ and $\delta_{\overline{q}}$. No external scaling matrix is allowed to affect the feedback signals between $\overline{P}$ and Δ because Δ is defined with a unity norm bound. Computing μ of the plant $\overline{P}$ with an additional scaling on the lower loop signals would consider a scaled set of operators $\mathbf{\Delta}$ which does not accurately represent the uncertainty description. Therefore, $\overline{P}$ only scales the $\delta_{\overline{q}}$ feedback signals.

Theorem 8.1.3 can be modified to compute a nominal flutter margin by changing the identify matrix in the scaling used to compute $\overline{P}$ to a zero matrix. This eliminates the feedback interconnection between the model and the uncertainty description $\mathbf{\Delta}$ so μ considers only the nominal dynamics and computes the smallest destabilizing perturbation to dynamic pressure and would result in $\Gamma_{rob} = \Gamma_{nom}$ for this simplified situation.

Including the uncertainty description $\mathbf{\Delta}$ ensures the robust flutter margin will be no greater than the nominal flutter margin. The robust flutter margin considers the model used to compute the nominal flutter margin which corresponds to the uncertainty operator $\Delta = 0 \in \mathbf{\Delta}$ and also the models corresponding to the remaining operators $\Delta \in \mathbf{\Delta}$. The conservatism in the robust flutter margin makes intuitive sense since since the nominal flutter margin is the worst-case stability boundary for a single model and the robust flutter margin is the worst-case stability boundary for a family of models.

$$\Gamma_{rob} \leq \Gamma_{nom} \tag{8.5}$$

The proof demonstrating the necessary and sufficient condition $\mu(\overline{P}) = 1$ also makes intuitive sense since the equality sign is needed to ensure the flutter margin is valid without being overly conservative. If $\mu(\overline{P}) < 1$ then there is no $\Delta \in \mathbf{\Delta}$ causing an instability and the flutter margin is too conservative. If $\mu(\overline{P}) > 1$ then the system is not robust to all modeling errors $\Delta \in \mathbf{\Delta}$ and the flutter margin is not a valid robust flutter margin.

Theorem 8.1.3 demonstrates a robust flutter margin can be computed by determining a scaling matrix $W_{\overline{q}}$ for which $\mu(\overline{P}) = 1$. Algorithm 8.1.4 implements an iterative approach to compute a scaling matrix for which $\mu(\overline{P}) \in (1 \pm \epsilon)$ for some desired level of accuracy ϵ.

Algorithm 8.1.4 *(robust flutter margin)*

Given
- *plant P at dynamic pressure $\overline{q}_o$*
- *operator $\delta_{\overline{q}}$ related to P as in Figure 5.2*
- *operator set $\mathbf{\Delta}$ with unity norm bound*

Define
- *weighting W_q to scale $\overline{q}$ feedback*
- *scalar $\epsilon > 0$ for accuracy*

Then

$$\overline{P} = P \begin{bmatrix} W_{\overline{q}} & 0 \\ 0 & I \end{bmatrix}$$

while $(\mu(\overline{P}) > 1 + \epsilon)$ OR $(\mu(\overline{P}) < 1 - \epsilon)$ {

$$W_{\overline{q}} = W_{\overline{q}} / \mu(\overline{P})$$

$$\overline{P} = P \begin{bmatrix} W_{\overline{q}} & 0 \\ 0 & I \end{bmatrix}$$

}

$\overline{q}_{rob} = \overline{q}_o + W_{\overline{q}}$

$\Gamma_{rob} = W_{\overline{q}}$

The dynamic pressure $\overline{q}_o$ defining the flight condition for the nominal plant dynamics must be chosen carefully to ensure the robust flutter margin computed with Algorithm 8.1.4 is valid. The nature of μ and the upper bound is such that all norm bounded operators centered around the origin are assumed to be valid perturbations to the plant dynamics so positive and negative perturbations to dynamic pressure are considered by the robust stability analysis. Thus, the robust flutter margin computed by $\mu(\overline{P}) = 1$ could correspond to either perturbation $\delta_{\overline{q}} = 1$ or $\delta_{\overline{q}} = -1$.

The actual dynamic pressure at the flutter instability is $\overline{q}_{rob} = \overline{q}_o - W_{\overline{q}}$ if $\delta_{\overline{q}} = -1$. A large weighting of $W_{\overline{q}} > \overline{q}_o$ indicates the flutter occurs at a negative dynamic pressure which may be unrealistic for classical flutter analysis. The value of the nominal dynamic pressure $\overline{q}_o$ may slide along the real axis to a large value without loss of generality in the μ analysis since this parameter linearly affects the nominal dynamics. A simple approach to ensure the robust flutter pressure is a positive dynamic pressure is demonstrated in Algorithm 8.1.5 which iterates over increasing values of $\overline{q}_o$ until the scaling associated with the robust flutter margin satisfies $W_{\overline{q}} < \overline{q}_o$.

Algorithm 8.1.5 *(robust flutter margin with $\overline{q}_o$ iteration)*
Given • *parameters as in Algorithm 8.1.4*
• *initial $\overline{q}_o$*
Then

```
valid_margin = FALSE
while (valid_margin == FALSE) {
        compute plant P at nominal dynamic pressure q̄_o
        compute q̄_rob and associated W_q̄ from Algorithm 8.1.4
         if (W_q̄ > q̄_o) then   q̄_o = 1.1 * W_q̄
         else                   valid_margin = TRUE
}
```

The exclusion of low and negative dynamic pressures for stability analysis may not be desirable for all applications related to aeroelasticity. An example of such an application is the analysis of aeroservoelastic dynamics for a high angle of attack aircraft which concerns instabilities at low dynamic pressures [78]. Algorithm 8.1.5 can be modified for these types of analysis by choosing a low value of $\overline{q}_o$ and find the scaling $W_{\overline{q}}$ corresponding to $\delta_{\overline{q}} = -1$ which computes the low dynamic pressure instability.

The flutter margin computation must allow for an arbitrary structure of operators $\overline{\mathbf{\Delta}}$ so an upper bound such as the function derived in Appendix A.3 must be used. Algorithms 8.1.4 and 8.1.5 can be adapted by replacing the μ calculation with a μ upper bound calculation to compute flutter margins. A search over frequency points is required when using the upper bound presented in Appendix A.3 so the accuracy of the robust flutter margin requires a dense grid of frequencies associated with the natural frequencies of the worst-case dynamics to be considered.

A simple approach may be implemented if the natural frequency of the unstable dynamics at the nominal flutter pressure can be assumed to be similar to the natural frequency of the unstable dynamics at the robust flutter pressure. This assumption can be justified if the uncertainty does not change the critical flutter mode between the nominal and robust flutter pressures which is often true for systems for relatively small levels of uncertainty and clear separation between critical and subcritical modal frequencies. Algorithm 8.1.6 presents this approach which first computes the frequency of the nominal flutter mode then computes a robust flutter margin from the μ upper bound evaluated near that frequency.

Algorithm 8.1.6 *(robust flutter margin with reduced frequency grid)*
Given • *parameters as in Algorithm 8.1.5*
Then

Compute $\overline{q}_{nom}$ using Algorithm 8.1.2
Compute frequency ω associated with $\overline{q}_{nom}$
Define dense frequency grid Ω centered around ω
Compute $\overline{q}_{rob}$ using Algorithm 8.1.5
(evaluate μ upper bound at Ω)

It is important to note a large and dense frequency grid increases the confidence that computed robustness measure is actually an upper bound for μ. Algorithm 8.1.6 must be used with caution since the assumptions behind its use may not be satisfied. It is recommended to compute a robust flutter margin with a dense frequency grid for a particular aircraft at several Mach numbers and compare the frequencies of the flutter dynamics to those of the nominal flutter dynamics. If these frequencies are similar, then Algorithm 8.1.6 may be considered for further analysis at different Mach numbers.

8.2 Robust Aeroservoelastic Stability Margins

Aeroservoelastic stability margins differ from flutter margins; however, the two concepts are conceptually similar. Essentially, a flutter margin relates the largest change in dynamic pressure that may be considered before an aeroelastic instability may be encountered by the open-loop plant. Similarly, an aeroservoelastic stability margin relates the largest change in dynamic pressure that may considered before an aeroservoelastic instability may be encountered by the closed-loop plant.

The similarity between these types of stability margins is a direct result of the models used in the analysis. The aeroelastic plant is formulated by relating the structural dynamics with the unsteady aerodynamics and then parameterizing the resulting system around dynamic pressure. Similarly, the aeroservoelastic plant is formulated by relating the structural dynamics with the unsteady aerodynamics and a control system and then parameterizing the resulting system around dynamic pressure. The relationship is actually close enough, as noted in Section 5.4, that an aeroservoelastic plant can be formulated by simply associating an aeroelastic plant with a control system.

The result of the similarity between aeroservoelastic and aeroelastic plans is that their respective stability margins can be computed using the same algorithms. The only difference, of course, is that the flutter margins are computed by analyzing the open-loop plant whereas the aeroservoelastic margins are computed by analyzing the closed-loop plant.

Nominal aeroservoelastic stability margins that determine the range of dynamic pressures over which the theoretical model is safe can thus be determined by applying Lemma 8.1.1 to the closed-loop aeroservoelastic plant. This plant can also be used to apply Algorithm 8.1.2 and compute the nominal aeroservoelastic stability margins.

Similarly, robust aeroservoelastic margins are formulated by directly applying the principles of robust flutter margins except the open-loop plant used for flutter analysis is replaced with the closed-loop plant. Thus, the robust aeroservoelastic stability margin is determined by applying Theorem 8.1.3 to the closed-loop aeroservoelastic plant with uncertainties and correspondingly computed with Algorithm 8.1.4.

Aeroservoelastic stability margins and flutter margins are similar as a result of the similarity in modeling concepts; however, the ranges of dynamic pressures that must be considered when evaluating stability are different. Notably, flutter is typically noted as an instability that is encountered at the edges of a flight envelope corresponding to high dynamic pressures but aeroservoelastic instabilities can be encountered at edges corresponding to high and low dynamic pressures.

The additional analysis of stability at low dynamic pressures requires an extension be developed for Algorithm 8.1.2 to ensure an accurate set of aeroservoelastic stability margins is computed. This extension is relatively straightforward to implement since it only changes the search direction along perturbations to dynamic pressure and is presented in Algorithm 8.2.1.

Algorithm 8.2.1 *(extended nominal aeroservoelastic stability margin)*

Given
- *closed-loop plant P at dynamic pressure $\overline{q}_o$*
- *operator $\delta_{\overline{q}}$ related to P as in Figure 5.1*

Define
- *scalars $\delta_{upper} > \delta_{lower}$ to bound $\delta_{\overline{q}}$*
- *scalar $\epsilon > 0$ for accuracy*

Then

while $(\delta_{upper} - \delta_{lower} > \epsilon)$ {

  $\delta_{\overline{q}} = \frac{1}{2}(\delta_{upper} + \delta_{lower})$

  if $(F_u(P, \delta_{\overline{q}})$ *has an unstable pole) then* $\delta_{lower} = \delta_{\overline{q}}$

  else $\delta_{upper} = \delta_{\overline{q}}$

}

$\overline{q}_{nom} = \overline{q}_o + \delta_{lower}$

$\Gamma_{nom} = \delta_{lower}$

The nominal aeroservoelastic stability margins are thus computed by considering additions to dynamic pressure as indicated by Algorithm 8.1.2 and also subtractions to dynamic pressure as indicated by Algorithm 8.2.1.

An extension must also be developed to compute robust aeroservoelastic stability margins; however, this extension is not as straightforward to implements as Algorithm 8.2.1. The additional complexity arise because robust aeroservoelastic stability margins are computed by analyzing the value of μ which considers norm bounded operators whereas nominal margins are computed by a simple search over scalar operators. Thus, the basic algorithm of Algorithm 8.1.4 does not immediately distinguish between positive or negative perturbations to dynamic pressure.

The issue of computing robust aeroservoelastic stability margins for the edge of the flight envelope that considers low dynamic pressures can be solved by considering the weighting introduced in Algorithm 8.1.5. This algorithm iterated over a weighting to the nominal dynamic pressure around which the plant to consistently increase the value until the flutter margin was less than that nominal dynamic pressure. This iteration ensured that no negative values of dynamic pressure were analyzed as the flutter margin. An extension to this approach that reduces, rather than increases, the dynamic pressure is presented in Algorithm 8.2.2.

Algorithm 8.2.2 *(extended robust aeroservoelastic stability margin)*

Given • *parameters as in Algorithm 8.1.4*
• *closed-loop P*
• *high value of initial $\overline{q}_o$*

Define • *scalar $\alpha > 0$*

Then

valid_margin = FALSE
while (valid_margin == FALSE*)* {
 compute plant P at nominal dynamic pressure $\overline{q}_o$
 compute $\overline{q}_{rob}$ from Algorithm 8.1.4
 if $(\overline{q}_{rob} > \overline{q}_o)$ then $\overline{q}_o = \overline{q}_o \quad \alpha$
 else *valid_margin* = TRUE
}

Algorithm 8.2.2 does not actually iterate based on a condition involving the scaling of dynamic pressure like is done in Algorithm 8.1.5; rather, the iteration is based on a condition for the dynamic pressure at which the instability occurs. This change in iteration condition is required because the Algorithm 8.1.5 was intended to ensure that negative values of dynamic pressure were not encountered as flutter margin but actually the aeroservoelastic stability margins may correspond to either positive or negative values of dynamic pressure.

The robust aeroservoelastic stability margins that are computed using Algorithm 8.2.2 attempt to find the low $\overline{q}$ edge of a flight envelope; however, a careful application of this algorithm is still required. Particularly, the initial value of nominal dynamic pressure around which the model is generated should be chosen to be high. The goal of this choice is to ensure that the algorithm does not begin by considering a dynamic pressure that is already lower than the instability pressure. If the initial value is higher than the actual instability pressure, then the algorithm will consistently reduce the value under consideration and should converge on the true stability margin.

Also, it should be noted that the stability margins that are computed by analyzing the closed-loop model may actually correspond to the flutter margins that were computed by analyzing the open-loop model. This possibility results because the aeroservoelastic model encompasses the aeroelastic model so they may share the same instabilities. Essentially, if the aeroservoelastic stability margins are identical to the flutter margins then the control system neither causes any instability nor alleviates any instability associated with the elastic dynamics of the aircraft.

8.3 Properties of Robust Stability Margins

Robust stability margins for both aeroelastic and aeroservoelastic analysis use μ as the worst-case stability parameter. μ is a much more informative stability margin as compared to traditional traditional parameters such as pole location and modal damping so it presents several advantages as an analysis parameter.

The conservatism introduced by considering the worst-case uncertainty perturbation can be interpreted as a measure of sensitivity. Robust μ values which are significantly different than the nominal stability margins indicate the plant is highly sensitive to modeling errors and changes in flight condition. A small perturbation to the system can drastically alter the aeroservoelastic stability properties. Conversely, similarity between the robust and nominal stability margins indicates the aircraft is not highly sensitive to small perturbations.

Robustness analysis determines not only the norm of the smallest destabilizing perturbation but also the direction. This information relates exact perturbations for which the system is particularly sensitive. μ can thus indicate the worst-case flutter mechanism which may naturally extend to indicate active and passive control strategies for flutter suppression. Similarly, μ indicates the mechanism causing an aeroservoelastic instability and may be used to indicate strategies for altering or augmenting the control system to eliminate the critical closed-loop instability.

Damping is only truly informative at the point of instability since stable damping at a given flight condition does not necessarily indicate an increase in dynamic pressure will be a stable flight condition [146]. μ computes the smallest destabilizing perturbation which indicates the nearest flight conditions that will cause a flutter instability. In this respect, μ is a stability predictor while damping is merely a stability indicator.

These characteristics of μ make the worst-case aeroservoelastic analysis especially valuable for flight test programs. Aeroelastic flight data can be measured at a stable flight condition and used to evaluate uncertainty operators. The μ method, unlike damping estimation, does not require the aircraft to approach instability for accurate prediction. μ can be computed to update the stability margins with respect to the new uncertainty levels. The worst-case stability margin then indicates what flight conditions may be safely considered.

Safe and efficient expansion of the flight envelope can be performed using an on-line implementation of the worst-case stability estimation algorithm. Computing μ does not introduce an excessive computational burden since the robust margins can often be analyzed in a similar amount of time as the estimates of damping parameters using standard off-the-shelf hardware and software packages. The predictive nature of μ and the computational efficiency allow a flutterometer tool to be developed that tracks the flutter margin during a flight test [76].

CHAPTER 9

μ METHOD

9.1 Model Updating

Generating a model by analyzing flight data is essential for computing a confident stability analysis. A nominal model generated purely from analytical equations of the predicted aircraft dynamics may not accurately describe the true aircraft; therefore, a model must be generated that accounts for the flight data to ensure the predicted dynamics represent the true dynamics.

The most direct method of generating a model from the flight data is to identify a system model entirely from the data measurements. Many system identification algorithms exist which have become standard tools for systems and control engineers [84]. Direct application of these methods to aeroelastic systems rarely produces an accurate model that completely represents the dynamics of the aircraft [66]. Aeroelastic response data is typically of poor quality relative to ground vibration test data because of low signal to noise ratio and unobserved dynamics in the response measurements; thus, system identification algorithms may be significantly less effective at estimating aeroelastic models as compared to estimating rigid-body models.

An alternative method is to utilize the nominal aircraft dynamical model as an initial estimate to model the true aircraft. The flight data is then used to update the elements of this model. Several methods have been devised to update an analytical structural model using experimental data [17]. Model updating can be performed on the full stress model or a subset computed with Guyan reduction [58]. Generally considering the full model is preferable since the reduction may distribute local errors throughout the entire model if an orthogonality condition is violated [71].

Two basic methods are proposed to update the full structural model using comparisons between experimental and predicted data. One method updates the mass and stiffness matrices of the finite element model [11]. This method suffers from lack of physical interpretation of the matrix updates and possible numerical conditioning. Another method updates specific parameters in the model [123]. This method is accurate for small systems but may require an excessive computational cost for large systems.

Aeroservoelastic models have an additional freedom to update the aerodynamic elements along with the structural elements. A method of updating the linear model in a modern control framework is developed [51]. This method may be overly conservative for describing nonlinearities and the corresponding stability margins are only accurate for flight conditions near the instability. A parametric identification algorithm is also developed that uses flight data to update specific terms in the aerodynamic model via a nonlinear optimization [36]. This method is sensitive to data anomalies and may show large variations variations in the resulting stability margins.

The approach taken to develop a μ method is to update only the uncertainty operators of the robust aeroelastic model using the flight data while leaving the nominal dynamics model unchanged. The model validation condition of Theorem 7.1.2 is used with the nominal plant P to generate a reasonable uncertainty description Δ to associate with P. Figure 9.1 describes the flow of information through this μ method.

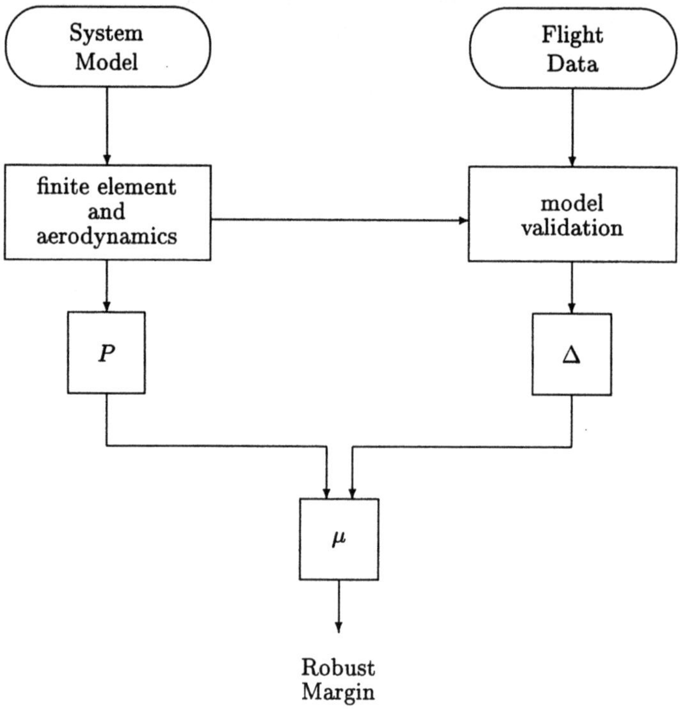

Fig. 9.1: Information Flowchart to Generate Plant and Uncertainty Operators from a System Model and Flight Data with the μ Method

Algorithm 9.1.1 presents a procedure to implement the μ method in a manner corresponding to Figure 9.1. The model update procedure is based on Algorithm 7.2.1 which actually scales the plant but has the equivalent effect of scaling the norm bound of the uncertainty description. The algorithm loops over the model validation procedure until an uncertainty description is determined that is not invalidated by the flight data. The final step is to compute a robust stability margin from the scaled plant by using a $\mu < 1$ condition as in Algorithm 8.1.6 for an aeroelastic or aeroservoelastic model.

Algorithm 9.1.1 *(robust flutter margins with model updating)*

Given • *plant P parametrized around $\overline{q}_o$*
• *uncertainty set $\boldsymbol{\Delta}$ associated with P*
• *input excitation data u*
• *output response data y*

Define • *matrix $W \geq I$ as the scaling update*

Then

while $(F_u(P, \Delta)$ *is invalidated by* u, y *using Algorithm 7.2.1)* {
$\boldsymbol{\Delta} = W\boldsymbol{\Delta}$
}
Γ_{rob} *is a stability margin computed from Algorithm 8.1.6*

There are several advantages to using this method as compared to traditional model updating methods. The typically poor quality of flight data in association with aircraft dynamics consisting of many modes makes updating a nominal model difficult. Traditional methods of norm-based update algorithms often generate a nonunique set of model updates with no way to determine which update has the most logical physical interpretation. The method of updating the uncertainty operators based on a worst-case magnitude avoids this problem.

Also, this method can work with flight data of varying quality. The updated uncertainty description will be highly accurate if the data shows a high signal to noise ratio and much of the dynamics are observed by the sensors. If the data does not have these desired characteristics, however, the method can still compute a flutter margin. An uncertainty description may be difficult to compute if the data does not indicate the aircraft dynamics well so the model validation procedure will not require a large magnitude for the uncertainty operators. The robust model in this situation will closely resemble the nominal model. In this way, the μ method will always generate a more accurate flutter margin and at worst the robust μ stability margin will be equivalent to the nominal stability margin.

9.2 Approaches to Utilize Flight Data

A flight test will generally consist of maneuvers at several test points which may be at identical or different flight conditions. The entire flight test program will utilize many flight tests to measure response data at test points throughout the flight envelope. The model updating method to generate uncertainty operators can use any or all of the entire set of flight data from the different test points.

Several approaches are formulated to utilize multiple flight data sets to update the uncertainty description associated with a nominal plant model. The uncertainty description may be different for each approach and the resulting flutter margin will be different for each approach due to the dependence of μ on the uncertainty set. Two approaches discussed here are denoted as local and global.

A local approach utilizes flight data from test points at identical flight conditions. This data is used to generate an uncertainty description for the nominal model at the particular flight condition associated with the data. The magnitude of the uncertainty operators is chosen such that robust model at the single flight condition is not invalidated by any of the flight data sets measured at that same flight condition with no consideration of data from other flight conditions.

The local approach presents the benefit of independently computing uncertainty descriptions for models at different flight conditions. This allows each uncertainty description to be more accurate since, for example, the flight data may indicate much smaller uncertainty operators are required for subsonic plant models even though large uncertainty operators are required for transonic plant models. The resulting worst-case stability margins will be less conservative since less uncertainty is required for the model.

Algorithm 9.2.1 outlines the local approach to utilize flight data.

Algorithm 9.2.1 *(local approach)*

Given
- *scalar* n_c *as number of flight conditions*
- *scalar* n_i *as number of data sets at condition* i
- *input data* $\{u_i^1, u_i^2, \ldots, u_i^{n_i}\}$ *at condition* i
- *output data* $\{y_i^2, y_i^2, \ldots, y_i^{n_i}\}$ *at condition* i
- *models* $\{P_1, P_2, \ldots, P_{n_c}\}$ *at condition* i

Then

```
for i = 1 : n_c {
        Choose initial Δ_i
        for j = 1 : n_i {
                validate F_u(P_i, Δ_i) using y_i^j and u_i^j
                increase size of Δ_i if necessary to validate
        }
        compute Γ_rob from μ(F_u(P_i, Δ_i))
}
```

A global approach utilizes the entire set of flight data from test points throughout the flight envelope to generate a single uncertainty description for all nominal aircraft models. The magnitudes of the uncertainty operators are chosen such that all nominal models with the associated uncertainty description are not invalidated by any of the flight data sets.

There are several advantages and disadvantages to using the global approach as compared to the local approach. A disadvantage is a possible large increase in conservatism of the robust stability margin since the uncertainty description is not minimized at each flight condition. A single particularly inaccurate plant model will require large uncertainty operators that may be highly conservative for plant models at flight conditions which are better representations of the true dynamics.

The advantage to this approach, however, is that the uncertainty description is truly worst-case with respect to the entire flight envelope. The worst-case errors from the worst-case flight condition are used to generate the uncertainty description for all conditions. Also, this approach is less sensitive to poorly measured flight data. A poorly modeled modal response may only appear in certain data sets. The local approach would not include uncertainty for these dynamics at conditions which did not clearly observe this modal response so the resulting stability margin would not account for the true level of modeling errors. Therefore, the flutter margin generated with the global approach may be more conservative but this introduces a corresponding higher margin of safety.

Algorithm 9.2.2 outlines the global approach to utilize flight data and compute robust stability margins for flutter or aeroservoelastic models.

Algorithm 9.2.2 *(global approach)*

Given
- *scalar n_c as number of flight conditions*
- *scalar n_i as number of data sets at condition i*
- *input data $\{u_i^1, u_i^2, \ldots, u_i^{n_i}\}$ at condition i*
- *output data $\{y_i^2, y_i^2, \ldots, y_i^{n_i}\}$ at condition i*
- *models $\{P_1, P_2, \ldots, P_{n_c}\}$ at condition i*

Then

Choose initial Δ
for $i = 1 : n_c$ {
    for $j = 1 : n_i$ {
        validate $F_u(P_i, \Delta)$ using y_i^j and u_i^j
        increase size of Δ if necessary to validate
    }
}
for $i = 1 : n_c$ {
    compute Γ_{rob} from $\mu\left(F_u(P_i, \Delta)\right)$
}

Hybrid approaches are also formulated that mix the local and global approaches. One straightforward hybrid approach would be to generate an uncertainty description using all data from a small range of flight conditions. This approach may be useful for separately considering sets of plant models which are generated using different techniques. For example, the model generating package called STARS computes all subsonic plant models with a doublet lattice algorithm while the supersonic models are generated with constant panel algorithms. A hybrid approach could be used to reflect this knowledge and consider groups of subsonic, supersonic and transonic plants independently.

The approaches outlined here are certainly not exhaustive. For example, a weighted norm approach can be formulated that uses flight data from the entire flight envelope but depends heavily on a particular subset of that data. Other approaches could concentrate on particular dynamics through modal filtering techniques to generate separate uncertainty description for individual modes [144].

CHAPTER 10
ROBUST STABILITY MARGINS OF A PITCH-PLUNGE SYSTEM

10.1 Equations of Motion

Robust aeroelastic and aeroservoelastic stability analysis is performed on a low-order system to demonstrate the procedure of applying the μ method. Experimental testing of this system differs from flight flutter testing in that the airspeed, not the dynamic pressure, is increased until an instability is encountered; however, the system can still be parameterized around dynamic pressure and associated stability margins computed as though the system represented an aircraft. Such an example is quite instructive because the equations of motion can be easily written so the derivation to formulate the model in the μ framework can be explicitly shown.

The system to be analyzed models an aeroservoelastic testbed that is operated by the Department of Aerospace Engineering at Texas A&M University. This testbed is a rigid airfoil attached to a support structure that allows pitch and plunge motion. The airfoil is mounted in a subsonic wind tunnel as shown in Figure 10.1.

Fig. 10.1: Aeroservoelastic Testbed at Texas A&M University

The support structure is composed of a carriage and a system of springs and cams. The carriage allows plunge motion by sliding on translational bearings while the springs and cams affect the angular pitch motion. An isometric drawing of the airfoil and support structure for the aeroelastic testbed is given in Figure 10.2.

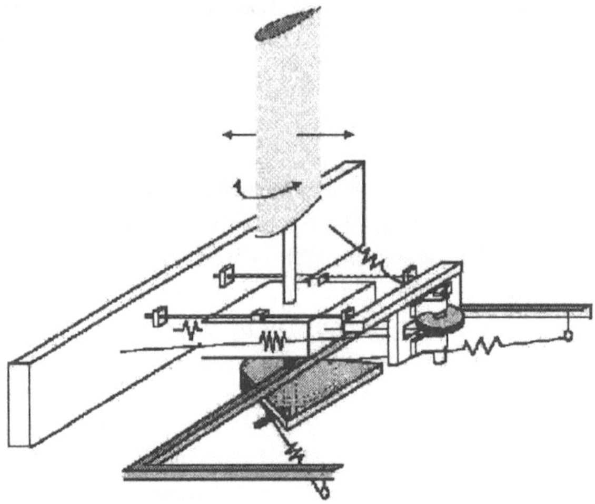

Fig. 10.2: Isometric Drawing of the Texas A&M University Aeroelastic Testbed

The testbed is developed to study behaviors related to classical bending and torsion mode interactions by equating the plunge motion with bending and the pitch motion with torsion. In particular, the open-loop characteristics of the system have been extensively studied by including different types of linear and nonlinear elements that affect the pitch dynamics [50, 96]. The inclusion of nonlinearities introduces several interesting behaviors such as limit cycle oscillations and internal resonances [97, 127]; however, the discussion of nonlinear aeroelasticity requires analysis procedures that range beyond the intended scope of this text [33, 35].

The testbed airfoil also has a flap that can be used as a control surface. The flap presents an additional degree of freedom for the system and may introduce instabilities such as control surface flutter and buzz; however, the stability analysis will only consider the interactions between structural dynamic modes for simplicity of presentation.

Equations of motion to describe the pitch and plunge motion during an aeroelastic response are derived from force and moment equations [67, 143]. These equations may be represented in matrix form.

$$\begin{bmatrix} m & mx_\alpha b \\ mx_\alpha b & I_\alpha \end{bmatrix} \begin{bmatrix} \ddot{y} \\ \ddot{\alpha} \end{bmatrix} + \begin{bmatrix} c_y & 0 \\ 0 & c_\alpha \end{bmatrix} \begin{bmatrix} \dot{y} \\ \dot{\alpha} \end{bmatrix} + \begin{bmatrix} k_y & 0 \\ 0 & k_\alpha \end{bmatrix} \begin{bmatrix} y \\ \alpha \end{bmatrix} = \begin{bmatrix} -L \\ M \end{bmatrix} \tag{10.1}$$

The element y is the plunge position and α is the pitch angle. Other variables include the nondimensional distance between the elastic axis and the center of mass x_α, the mass of the wing m, the mass moment of inertia I_α, semichord length b, structural damping coefficients in pitch and plunge c_y and c_α, and spring constants k_y and k_α.

The lift L and moment M are determined by quasi-steady aerodynamic theory. The parameters c_{l_α} and c_{m_α} are introduced to represent the lift and moment coefficients for angle of attack and c_{l_β} and c_{m_β} are introduced to represent the lift and moment coefficients for control surface position.

$$L = \bar{q}2bc_{l_\alpha}\left(\alpha + \left(\frac{1}{2} - e\right)\frac{b}{U}\dot{\alpha} + \frac{1}{U}\dot{y}\right) + \bar{q}bc_{l_\beta}\beta \tag{10.2}$$

and

$$M = \bar{q}2b^2c_{m_\alpha}\left(\alpha + \left(\frac{1}{2} - e\right)\frac{b}{U}\dot{\alpha} + \frac{1}{U}\dot{y}\right) + \bar{q}b^2c_{m_\beta}\beta \tag{10.3}$$

The stability analysis is limited to considering the dynamics of the model when the tunnel airspeed is set at $U = 6\ m/s$. A more extensive analysis can certainly be performed; however, analysis of a single flight condition is sufficient for this example.

Values of the model parameters are given in Table 10.1.

Table 10.1: Nominal Parameters for Model of Pitch-Plunge System

$U = 6\ m/s$
e = -0.6
$m = 12.387\ kg$
$b = 0.135\ m$
$I_\alpha = 0.065\ kg\ m^2$
$x_\alpha = .2466$
$k_y = 2844.4\ N/m$
$k_\alpha = 3.525\ Nm/rad$
$c_y = 27.43\ kg/s$
$c_\alpha = 0.036\ kg\ m^2/s$
$c_{l_\alpha} = 6.28$
c_{m_α} = -0.635
$c_{l_\beta} = 3.358$
$c_{m_\beta} = 12.387$

Free-decay responses are measured with this system by setting the support system to a desired pitch angle and plunge displacement and then releasing. These responses are simulated using a model by selecting initial conditions to match the support system settings.

10.2 Nominal Aeroelastic Model

The equations of motion must be parameterized over a flight condition to introduce a perturbation that will relate the stability margin. Dynamic pressure and airspeed are the flight condition parameters and these affect the system through the lift and moment equations. These equations are linear functions of dynamic pressure and nonlinear functions of airspeed; therefore, the lift and moment equations will be parameterized around a perturbation to dynamic pressure in order to compute stability margins.

Define dynamic pressure as a nominal value, $\overline{q}_o$, and a perturbation, $\delta_{\overline{q}} \in \mathbf{R}$, that affects the nominal value.

$$\overline{q} = \overline{q}_o + \delta_{\overline{q}} \tag{10.4}$$

Separate the effects of the perturbation from the equation that relates the structural response of the system to the aerodynamic lift force.

$$\begin{aligned}
& m\ddot{y} + mx_\alpha b\ddot{\alpha} + c_y\dot{y} + k_y y \\
&= -\overline{q}2\left(bc_{l_\alpha}\left(\alpha + \frac{\dot{y}}{U} + \left(\frac{1}{2} - e\right)b\frac{\dot{\alpha}}{U}\right) + bc_{l_\beta}\beta\right) \\
&= -\left(\overline{q}_o + \delta_{\overline{q}}\right)2\left(bc_{l_\alpha}\left(\alpha + \frac{\dot{y}}{U} + \left(\frac{1}{2} - e\right)b\frac{\dot{\alpha}}{U}\right) + bc_{l_\beta}\beta\right) \\
&= -\overline{q}_o2\left(bc_{l_\alpha}\left(\alpha + \frac{\dot{y}}{U} + \left(\frac{1}{2} - e\right)b\frac{\dot{\alpha}}{U}\right) + bc_{l_\beta}\beta\right) \\
&\qquad -\delta_{\overline{q}}2\left(bc_{l_\alpha}\left(\alpha + \frac{\dot{y}}{U} + \left(\frac{1}{2} - e\right)b\frac{\dot{\alpha}}{U}\right) + bc_{l_\beta}\beta\right) \\
&= -\overline{q}_o2\left(bc_{l_\alpha}\left(\alpha + \frac{\dot{y}}{U} + \left(\frac{1}{2} - e\right)b\frac{\dot{\alpha}}{U}\right) + bc_{l_\beta}\beta\right) - \delta_{\overline{q}}z_l \\
&= -\overline{q}_o2\left(bc_{l_\alpha}\left(\alpha + \frac{\dot{y}}{U} + \left(\frac{1}{2} - e\right)b\frac{\dot{\alpha}}{U}\right) + bc_{l_\beta}\beta\right) - w_l
\end{aligned} \tag{10.5}$$

This formulation eliminates the perturbation to dynamic pressure from the equation by replacing the terms involving $\delta_{\overline{q}}$ with a set of additional input and output signals w_l and z_l. These signals differ from traditional signals in that neither represents physical signals or measurements; instead, they are fictitious mathematical signals used for the purpose of theoretical analysis. The feedback relationship $w_l = \delta_{\overline{q}}z_l$ relates the perturbation and the nominal dynamics where z_l is a function of the states and control surface position.

$$\begin{aligned}
z_l &= 2\left(bc_{l_\alpha}\left(\alpha + \frac{\dot{y}}{U} + \left(\frac{1}{2} - e\right)b\frac{\dot{\alpha}}{U}\right) + bc_{l_\beta}\beta\right) \\
&= z_{l_\alpha}\alpha + z_{l_{\dot{y}}}\dot{y} + z_{l_{\dot{\alpha}}}\dot{\alpha} + z_{l_\beta}\beta
\end{aligned} \tag{10.6}$$

A similar derivation is performed to eliminate the effects of $\delta_{\overline{q}}$ from the equation that relates the structural response of the system to the aerodynamic moment.

$$
\begin{aligned}
& m x_\alpha b \ddot{y} + I_\alpha \ddot{\alpha} + c_\alpha \alpha + k_\alpha \alpha \\
& = \overline{q} 2 \left(b^2 c_{m_\alpha} \left(\alpha + \frac{\dot{y}}{U} + \left(\frac{1}{2} - e \right) b \frac{\dot{\alpha}}{U} \right) + b^2 c_{m_\beta} \beta \right) \\
& = (\overline{q}_o + \delta_{\overline{q}}) 2 \left(b^2 c_{m_\alpha} \left(\alpha + \frac{\dot{y}}{U} + \left(\frac{1}{2} - e \right) b \frac{\dot{\alpha}}{U} \right) + b^2 c_{m_\beta} \beta \right) \\
& = \overline{q}_o 2 \left(b^2 c_{m_\alpha} \left(\alpha + \frac{\dot{y}}{U} + \left(\frac{1}{2} - e \right) b \frac{\dot{\alpha}}{U} \right) + b^2 c_{m_\beta} \beta \right) \\
& \qquad + \delta_{\overline{q}} 2 \left(b^2 c_{m_\alpha} \left(\alpha + \frac{\dot{y}}{U} + \left(\frac{1}{2} - e \right) b \frac{\dot{\alpha}}{U} \right) + b^2 c_{m_\beta} \beta \right) \\
& = \overline{q}_o 2 \left(b^2 c_{m_\alpha} \left(\alpha + \frac{\dot{y}}{U} + \left(\frac{1}{2} - e \right) b \frac{\dot{\alpha}}{U} \right) + b^2 c_{m_\beta} \beta \right) + \delta_{\overline{q}} z_m \\
& = \overline{q}_o 2 \left(b^2 c_{m_\alpha} \left(\alpha + \frac{\dot{y}}{U} + \left(\frac{1}{2} - e \right) b \frac{\dot{\alpha}}{U} \right) + b^2 c_{m_\beta} \beta \right) + w_m
\end{aligned} \tag{10.7}
$$

A set of fictitious inputs and outputs is again introduced to the formulation to replace the explicit dependence on $\delta_{\overline{q}}$. The feedback relationship $w_m = \delta_{\overline{q}} z_m$ is used to affect the nominal dynamics with an output signal z_m that is defined as a function of the states and control surface position.

$$
\begin{aligned}
z_m & = 2 \left(b^2 c_{m_\alpha} \left(\alpha + \frac{\dot{y}}{U} + \left(\frac{1}{2} - e \right) b \frac{\dot{\alpha}}{U} \right) + b^2 c_{m_\beta} \beta \right) \\
& = z_{m_\alpha} \alpha + z_{m_{\dot{y}}} \dot{y} + z_{m_{\dot{\alpha}}} \dot{\alpha} + z_{m_\beta} \beta
\end{aligned} \tag{10.8}
$$

A state-space model of the pitch-plunge system can be formulated by combining the lift and moment equations of motion. The additional input and output signals must be preserved in the derivation to ensure the effect of the perturbation to dynamic pressure is identical for the original equations of motion and the state-space formulation. This derivation can be accomplished by scaling and subtracting (10.5) and (10.7). The states of the model are chosen to be the plunge displacement and rate, y and $\dot{y}$, and the pitch angle and rate, α and $\dot{\alpha}$.

Consider the equation involving $\ddot{y}$ that results by eliminating the explicit dependence on $\ddot{\alpha}$.

$$
\begin{aligned}
\ddot{y} = {} & \left(I_\alpha m - m^2 x_\alpha^2 b^2\right)^{-1} \\
& * \left[\left(-I_\alpha k_y\right) y + \left(m x_\alpha b k_\alpha - \overline{q}_o 2 \left(I_\alpha b c_{l_\alpha} + m x_\alpha b^3 c_{m_\alpha}\right)\right) \alpha \right. \\
& + \left(-I_\alpha c_y - \overline{q}_o 2 \left(I_\alpha b c_{l_\alpha} + m x_\alpha b^3 c_{m_\alpha}\right) \frac{1}{U}\right) \dot{y} \\
& + \left(m x_\alpha b c_\alpha - \overline{q}_o 2 \left(I_\alpha b c_{l_\alpha} + m x_\alpha b^3 c_{m_\alpha}\right) \left(\frac{1}{2} - e\right) b \frac{1}{U}\right) \dot{\alpha} \\
& + \left(-I_\alpha\right) w_l + \left(-m x_\alpha b\right) w_m \\
& \left. + \left(-\overline{q}_o 2 \left(I_\alpha b c_{l_\beta} + m x_\alpha b^3 c_{m_\beta}\right)\right) \beta \right] \\
= {} & a_{31} y + \left(a_{32_o} + a_{32_\alpha} k_\alpha\right) \alpha + \left(a_{33_o} + a_{33_y} c_y\right) \dot{y} + a_{34} \dot{\alpha} \\
& + b_{3l} w_l + b_{3m} w_m + b_{3\beta} \beta \\
= {} & a_{31} y + a_{32} \alpha + a_{33} \dot{y} + a_{34} \dot{\alpha} + b_{3l} w_l + b_{3m} w_m + b_{3\beta} \beta
\end{aligned} \tag{10.9}
$$

Equation 10.9 presents a simplified form that is convenient for displaying the state-space model. Terms are defined that combine the expressions dependent on the system parameters into scalars that are easier to display. The terms are named to agree with the standard nomenclature that uses A as the state matrix and B as the control effectiveness of the state-space model. Thus, the a terms denote coefficients that scale the state variables and are elements of the state matrix. Similarly, the b terms denote coefficients that scale the input signals and are elements of the control effectiveness matrix.

The other state equation results from combining (10.5) and (10.7) to eliminate terms involving $\ddot{y}$.

$$
\begin{aligned}
\ddot{\alpha} = {} & \left(I_\alpha - m x_\alpha^2 b^2\right)^{-1} \\
& * \left[\left(x_\alpha b k_y\right) y + \left(-k_\alpha + \overline{q}_o 2 \left(b^2 c_{m_\alpha} + x_\alpha b^2 c_{l_\alpha}\right)\right) \alpha \right. \\
& + \left(x_\alpha b c_y + \overline{q}_o 2 \left(b^2 c_{m_\alpha} + x_\alpha b^2 c_{l_\alpha}\right) \frac{1}{U}\right) \dot{y} \\
& + \left(-c_\alpha + \overline{q}_o 2 \left(b^2 c_{m_\alpha} + x_\alpha b^2 c_{l_\alpha}\right) \left(\frac{1}{2} - e\right) b \frac{1}{U}\right) \dot{\alpha} \\
& + \left(x_\alpha b\right) w_l + w_m \\
& \left. + \left(\overline{q}_o 2 \left(b^2 c_{m_\beta} + x_\alpha b^2 c_{l_\beta}\right)\right) \beta \right] \\
= {} & a_{41} y + \left(a_{42_o} + a_{42_\alpha} k_\alpha\right) \alpha + \left(a_{43_o} + a_{43_y} c_y\right) \dot{y} + a_{44} \dot{\alpha} \\
& + b_{4l} w_l + b_{4m} w_m + b_{4\beta} \beta \\
= {} & a_{41} y + a_{42} \alpha + a_{43} \dot{y} + a_{44} \dot{\alpha} + b_{4l} w_l + b_{4m} w_m + b_{4\beta} \beta
\end{aligned} \tag{10.10}
$$

Simplified elements for the state and control effectiveness matrices are again defined to simplify (10.10).

A state-space model of the pitch-plunge system that includes the feedback signals to relate the nominal dynamics and the perturbation to dynamic pressure can be formulated by combining the state-update equations of (10.9) and (10.10).

$$\begin{bmatrix} \dot{y} \\ \dot{\alpha} \\ \ddot{y} \\ \ddot{\alpha} \\ \hline z_l \\ z_m \\ y \\ \alpha \end{bmatrix} = \left[\begin{array}{cccc|ccc} 0 & 0 & 1 & 0 & 0 & 0 & 0 \\ 0 & 0 & 0 & 1 & 0 & 0 & 0 \\ a_{31} & a_{32} & a_{33} & a_{34} & b_{3l} & b_{3m} & b_{3\beta} \\ a_{41} & a_{42} & a_{43} & a_{44} & b_{4l} & b_{4m} & b_{4\beta} \\ \hline 0 & z_{l_\alpha} & z_{l_{\dot{y}}} & z_{l_{\dot{\alpha}}} & 0 & 0 & z_{l_\beta} \\ 0 & z_{m_\alpha} & z_{m_{\dot{y}}} & z_{m_{\dot{\alpha}}} & 0 & 0 & z_{m_\beta} \\ 1 & 0 & 0 & 0 & 0 & 0 & 0 \\ 0 & 1 & 0 & 0 & 0 & 0 & 0 \end{array}\right] \begin{bmatrix} y \\ \alpha \\ \dot{y} \\ \dot{\alpha} \\ \hline w_l \\ w_m \\ \beta \end{bmatrix} \tag{10.11}$$

This state-space model has 4 output signals and 3 input signals. The outputs are comprised of z_l and z_m for feedback to the dynamic pressure perturbation and y and α to represent sensor measurements. The inputs are comprised of w_l and w_m for feedback from the dynamic pressure perturbation and β to represent the control surface position. The sensor measurements and control surface positions are not explicitly needed for stability analysis of the open-loop aeroelastic system; however, they will be used for stability analysis of the closed-loop aeroservoelastic system and so are included here.

The block diagram of the open-loop state-space model, P, and the perturbation to dynamic pressure, $\delta_{\bar{q}}$, is shown in Figure 10.3.

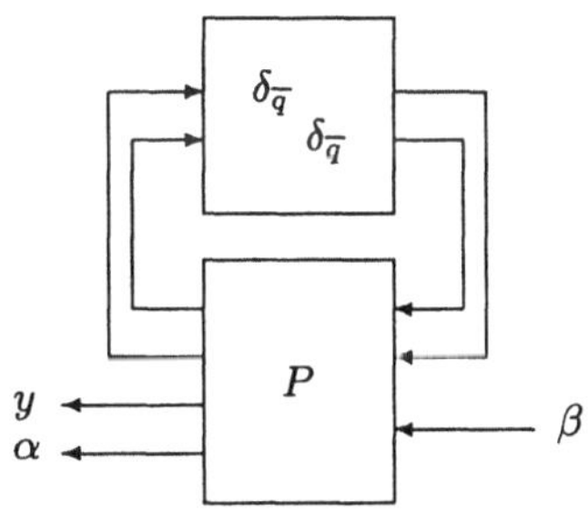

Fig. 10.3: Nominal Aeroelastic Model of Pitch-Plunge System with Parameterization over Dynamic Pressure and Uncertainty Description

The perturbation $\delta_{\bar{q}}$ actually appears twice in the feedback operator in Figure 10.3 to relate the pair of additional inputs and outputs. This is expected and agrees with the initial formulation of the μ method that noted the perturbation to dynamic pressure should appear once for every aeroelastic mode. Thus, the state-space model for the pitch-plunge system has two inputs and outputs that are related by the two instances of $\delta_{\overline{q}}$.

The modal properties of the pitch-plunge model are determined by analyzing the eigenvalues of the state matrix of (10.11). Consider the elements of this matrix when the parameters in Table 10.1 are substituted for the variables.

$$A = \begin{bmatrix} 0 & 0 & 1 & 0 \\ 0 & 0 & 0 & 1 \\ -291.112 & -1.206 & -3.390 & -0.063 \\ 1846.888 & -54.429 & -20.197 & -0.348 \end{bmatrix} \tag{10.12}$$

The eigenvalues of this model are given by $\lambda_1 = -0.43 \pm 8.00$ and $\lambda_2 = -1.44 \pm 16.71$. The associated natural frequencies and dampings for the modes of the nominal model are presented in Table 10.2.

Table 10.2: Modal Parameters for the Nominal Aeroelastic Pitch-Plunge Model

frequency (rad/s)	damping
8.016	.053
16.771	.086

10.3 Robust Aeroelastic Model

The nominal state-space model is only an approximation to the true dynamics of the pitch-plunge model so an uncertainty description must be developed to account for errors. An uncertainty description could be developed by carefully considering the equations of motion and noting possible deficiencies arising from the simplified quasi-steady aerodynamics and the linear assumptions of the structural dynamics; however, such a complicated uncertainty description is not required here. This tutorial example is meant to illustrate the procedure of including uncertainty to formulate a robust aeroelastic model so a relatively simple uncertainty description will be chosen.

Consider an additive perturbation associated with the structural damping parameter for the plunge displacement in the lift equation. The true structural damping, c_y, is assumed to be the nominal value of damping, c_{y_o}, plus some value of the perturbation, δ_{c_y}, that describes the error between the nominal value and the true value.

$$\begin{aligned} c_y &= c_{y_o} + W_{c_y}\delta_{c_y} \\ &= c_{y_o} + 5.0\delta_{c_y} \end{aligned} \tag{10.13}$$

A weighting, $W_{c_y} \in \mathbf{R}$, is associated with the uncertainty parameter to ensure a robust stability analysis using μ properly accounts for the anticipated levels of modeling uncertainty. Recall the structural singular value is defined based on the assumption that all uncertainty operators are norm bounded by 1. Thus, including the weighting of $W_{c_y} = 5.0$ implies the true value of structural damping is anticipated to be no more than 5 *kg/s* away from the nominal value of 27.43 *kg/s*.

A new equation of motion relating the structural response of the system to the aerodynamic lift force can be derived that includes the effects of this uncertainty. It should be noted that only the lift equation needs to be modified because the moment equation does not depend on c_y.

$$\begin{aligned} m\ddot{y} + mx_\alpha b\ddot{\alpha} + c_y\dot{y} + k_y y &= m\ddot{y} + mx_\alpha b\ddot{\alpha} + \left(c_{y_o} + W_{c_y}\delta_{c_y}\right)\dot{y} + k_y y \\ &= m\ddot{y} + mx_\alpha b\ddot{\alpha} + c_{y_o}\dot{y} + k_y y + \delta_{c_y} W_{c_y}\dot{y} \\ &= m\ddot{y} + mx_\alpha b\ddot{\alpha} + c_{y_o}\dot{y} + k_y y + \delta_{c_y} z_y \\ &= m\ddot{y} + mx_\alpha b\ddot{\alpha} + c_{y_o}\dot{y} + k_y y + w_y \end{aligned} \tag{10.14}$$

The explicit dependence of the lift equation on the uncertainty parameter associated with structural damping is eliminated in the new equation of motion by replacing the terms that depend on δ_{c_y} with the feedback signals w_y and z_y. These signals represent the fictitious input and output pair that is used to relate the nominal dynamics and the uncertainty in a feedback manner. The signals are related by $w_y = \delta_{c_y} z_y$ where z_y is the weighted value of plunge rate.

$$z_y = W_{c_y}\dot{y} \tag{10.15}$$

Consider also an additive perturbation to the stiffness associated with the pitch dynamics. The true value of this stiffness, k_α, is assumed to be the sum of the nominal value, k_{α_o}, and a perturbation, δ_{k_α}, that is represented by an uncertainty parameter.

$$\begin{aligned} k_\alpha &= k_{\alpha_o} + W_{k_\alpha}\delta_{k_\alpha} \\ &= k_{\alpha_o} + 0.35\delta_{k_\alpha} \end{aligned} \tag{10.16}$$

The uncertainty parameter is again weighted to ensure the values of δ_{k_α} between 1 and -1 that are considered by μ analysis are actually indicating the anticipated levels of potential modeling errors. Thus, the weighting of $W_{k_\alpha} = .35$ ensures the resulting stability margin will reflect the robustness of the model to variations in stiffness between $3.5 \pm .35$ *Nm/rad*.

Derive the new equation of motion that relates the structural dynamics and the aerodynamic moment using the uncertain value of pitch stiffness.

$$\begin{aligned}
mx_\alpha b\ddot{y} + I_\alpha\ddot{\alpha} + c_\alpha\dot{\alpha} + k_\alpha\alpha \\
&= mx_\alpha b\ddot{y} + I_\alpha\ddot{\alpha} + c_\alpha\dot{\alpha} + (k_{\alpha_o} + W_{k_\alpha}\delta_{k_\alpha})\,\alpha \\
&= mx_\alpha b\ddot{y} + I_\alpha\ddot{\alpha} + c_\alpha\dot{\alpha} + k_{\alpha_o}\alpha + \delta_{k_\alpha}W_{k_\alpha}\alpha \\
&= mx_\alpha b\ddot{y} + I_\alpha\ddot{\alpha} + c_\alpha\dot{\alpha} + k_{\alpha_o}\alpha + \delta_{k_\alpha}z_\alpha \\
&= mx_\alpha b\ddot{y} + I_\alpha\ddot{\alpha} + c_\alpha\dot{\alpha} + k_{\alpha_o}\alpha + w_\alpha
\end{aligned} \tag{10.17}$$

This new equation is similar to the new lift equation in the sense that both equations have eliminated the explicit dependence on any uncertainty parameters by introducing additional signals. This moment equation uses the relationship $w_\alpha = \delta_{k_\alpha}z_\alpha$ to relate the nominal dynamics and the uncertainty parameters in a feedback manner where z_α is the additional output of the scaled pitch angle.

$$z_\alpha = W_{k_\alpha}\alpha \tag{10.18}$$

A state-space realization of the equations of motion is formulated by equating the force equations of (10.2) and (10.14) to generate a new force equation. Similarly, the moment equations of (10.3) and (10.17) are equated to derive the second equation that describes the dynamics of the system. Also, the derivation for the state-space realization of the robust aeroelastic model must retain the additional signals that parameterize the system over dynamic pressure.

Scale and add the new force equations to derive an equation for $\ddot{y}$ that does not depend explicitly on $\ddot{\alpha}$.

$$\begin{aligned}
\ddot{y} = {} & \left(I_\alpha m - m^2x_\alpha^2b^2\right)^{-1} \\
& * \Big[\left(-I_\alpha k_y\right)y + \left(mx_\alpha bk_{\alpha_o} - \bar{q}_o 2\left(I_\alpha bc_{l_\alpha} + mx_\alpha b^3c_{m_\alpha}\right)\right)\alpha \\
& + \left(-I_\alpha c_{y_o} - \bar{q}_o 2\left(I_\alpha bc_{l_\alpha} + mx_\alpha b^3c_{m_\alpha}\right)\frac{1}{U}\right)\dot{y} \\
& + \left(mx_\alpha bc_\alpha - \bar{q}_o 2\left(I_\alpha bc_{l_\alpha} + mx_\alpha b^3c_{m_\alpha}\right)\left(\frac{1}{2} - e\right)b\frac{1}{U}\right)\dot{\alpha} \\
& + \left(-I_\alpha\right)w_l + \left(-mx_\alpha b\right)w_m + \left(-I_\alpha\right)w_y + \left(mx_\alpha b\right)w_\alpha \\
& + \left(-\bar{q}_o 2\left(I_\alpha bc_{l_\beta} + mx_\alpha b^3c_{m_\beta}\right)\right)\beta \Big] \\
= {} & a_{31}y + \left(a_{32_o} + a_{32_\alpha}k_{\alpha_o}\right)\alpha + \left(a_{33_o} + a_{33_y}c_{y_o}\right)\dot{y} + a_{34}\dot{\alpha} \\
& + b_{3l}w_l + b_{3m}w_m + b_{3y}w_y + b_{3\alpha}w_\alpha + b_{3\beta}\beta
\end{aligned} \tag{10.19}$$

The additional inputs of w_m and w_y are a significant difference between the equation for $\ddot{y}$ of the robust aeroelastic model in (10.19) and the corresponding $\ddot{y}$ equation of the nominal aeroelastic model in (10.9); however, there are many expected similarities that should be noted. One similarity is the relationship between the coefficients that scale the fictitious signals associated with uncertainty and parameterization over dynamic pressure. This relationship is expected because the signal w_y enters the equations in the same manner as w_l and the signal w_α enters the equations in the same manner, but with opposite sign, as w_m. Thus, their coefficients are related.

$$b_{3y} = b_{3l} \tag{10.20}$$

$$b_{3\alpha} = -b_{3m} \tag{10.21}$$

Another similarity between (10.19) and (10.9) is the exact match between the remaining terms such as a_{31} and b_{3l}. This match results because the only difference between the two derivations is the inclusion of parametric uncertainty and the effects of this uncertainty are restricted to the effects of the two additional input signals. Of course, different state-space realizations could be formulated by standard transformations so the match is not always guaranteed. Also, the weightings associated with each uncertainty are scalars so there are no additional states introduced in this model but certainly there are models that will require additional states to describe weighting functions.

The other state equation results from combining the force and moments equations to eliminate terms involving $\ddot{y}$.

$$\begin{aligned}
\ddot{\alpha} = {} & \left(I_\alpha - m x_\alpha^2 b^2\right)^{-1} \\
& * \Big[\left(x_\alpha b k_y\right) y + \left(-k_{\alpha_o} + \bar{q}_o 2\left(b^2 c_{m_\alpha} + x_\alpha b^2 c_{l_\alpha}\right)\right) \alpha \\
& + \left(x_\alpha b c_{y_o} + \bar{q}_o 2\left(b^2 c_{m_\alpha} + x_\alpha b^2 c_{l_\alpha}\right) \frac{1}{U}\right) \dot{y} \\
& + \left(-c_\alpha + \bar{q}_o 2\left(b^2 c_{m_\alpha} + x_\alpha b^2 c_{l_\alpha}\right) \left(\frac{1}{2} - e\right) b \frac{1}{U}\right) \dot{\alpha} \\
& + \left(x_\alpha b\right) w_l + w_m + \left(x_\alpha b\right) w_y - w_\alpha \\
& + \left(\bar{q}_o 2\left(b^2 c_{m_\beta} + x_\alpha b^2 c_{l_\beta}\right)\right) \beta \Big] \\
= {} & a_{41} y + \left(a_{42_o} + a_{42_\alpha} k_\alpha\right) \alpha + \left(a_{43_o} + a_{43_y} c_y\right) \dot{y} + a_{44} \dot{\alpha} \\
& + b_{4l} w_l + b_{4m} w_m + b_{4y} w_y + b_{4\alpha} w_\alpha + b_{4\beta} \beta
\end{aligned} \tag{10.22}$$

The same relationships exist between the equations for $\ddot{\alpha}$ in (10.22) and (10.10) as there were for the corresponding $\ddot{y}$ equations. Namely, they are identical excepting for the additional terms resulting from the additional signals w_y and w_α in the robust aeroelastic model. Also, the coefficients that scale these new input signals are related to the coefficients of the terms that scale the signals used to parameterize the model around dynamic pressure.

$$b_{4y} = b_{4l} \tag{10.23}$$
$$b_{4\alpha} = -b_{4m} \tag{10.24}$$

Equations (10.19) and (10.22) can be combined into matrix form to generate a state-space realization to model the pitch-plunge system.

$$\begin{bmatrix} \dot{y} \\ \dot{\alpha} \\ \ddot{y} \\ \ddot{\alpha} \\ \hline z_l \\ z_m \\ z_y \\ z_\alpha \\ y \\ \alpha \end{bmatrix} = \left[\begin{array}{c|c} A & B \\ \hline C & D \end{array}\right] \begin{bmatrix} y \\ \alpha \\ \dot{y} \\ \dot{\alpha} \\ \hline w_l \\ w_m \\ w_y \\ w_\alpha \\ \beta \end{bmatrix} \tag{10.25}$$

$$\begin{aligned} A &= \begin{bmatrix} 0 & 0 & 1 & 0 \\ 0 & 0 & 0 & 1 \\ a_{31} & a_{32_o} + a_{32_\alpha} k_{\alpha_o} & a_{33_o} + a_{33_y} c_{y_o} & a_{34} \\ a_{41} & a_{42_o} + a_{42_\alpha} k_{\alpha_o} & a_{43_o} + a_{43_y} c_{y_o} & a_{44} \end{bmatrix} \\ &= \begin{bmatrix} 0 & 0 & 1 & 0 \\ 0 & 0 & 0 & 1 \\ -291.112 & -1.206 & -3.390 & -0.063 \\ 1846.888 & -54.429 & -20.197 & -0.348 \end{bmatrix} \end{aligned} \tag{10.26}$$

$$\begin{aligned} B &= \begin{bmatrix} 0 & 0 & 0 & 0 & 0 \\ 0 & 0 & 0 & 0 & 0 \\ b_{3l} & b_{3m} & b_{3y} & b_{3\alpha} & b_{3\beta} \\ b_{4l} & b_{4m} & b_{4y} & b_{4\alpha} & b_{4\beta} \end{bmatrix} \\ &= \begin{bmatrix} 0 & 0 & 0 & 0 & 0 \\ 0 & 0 & 0 & 0 & 0 \\ -0.102 & -0.649 & -0.102 & 0.649 & -8.510 \\ 0.649 & 19.504 & 0.649 & -19.504 & 207.156 \end{bmatrix} \end{aligned} \tag{10.27}$$

$$C = \begin{bmatrix} 0 & z_{l_\alpha} & z_{l_{\dot{y}}} & z_{l_{\dot{\alpha}}} \\ 0 & z_{m_\alpha} & z_{m_{\dot{y}}} & z_{m_{\dot{\alpha}}} \\ 0 & 0 & W_{c_y} & 0 \\ W_{k_\alpha} & 0 & 0 & 0 \\ 1 & 0 & 0 & 0 \\ 0 & 1 & 0 & 0 \end{bmatrix}$$

$$= \begin{bmatrix} 0 & 1.695 & 0.282 & 0.042 \\ 0 & -0.023 & -0.004 & -0.001 \\ 0 & 0 & 5 & \\ 0 & 35 & 0 & 0 \\ 1 & 0 & 0 & 0 \\ 0 & 1 & 0 & 0 \end{bmatrix} \tag{10.28}$$

$$D = \begin{bmatrix} 0\,0\,0\,0 & z_{l_\beta} \\ 0\,0\,0\,0 & z_{m_\beta} \\ 0\,0\,0\,0 & 0 \\ 0\,0\,0\,0 & 0 \\ 0\,0\,0\,0 & 0 \\ 0\,0\,0\,0 & 0 \end{bmatrix} = \begin{bmatrix} 0\,0\,0\,0 & 0.907 \\ 0\,0\,0\,0 & 0.452 \\ 0\,0\,0\,0 & 0 \\ 0\,0\,0\,0 & 0 \\ 0\,0\,0\,0 & 0 \\ 0\,0\,0\,0 & 0 \end{bmatrix} \tag{10.29}$$

The state-space model of the robust aeroelastic system has 6 output signals and 5 input signals. The outputs are comprised of z_l and z_m for feedback to the dynamic pressure perturbation, z_y and z_α for feedback to the uncertainty operators, and y and α to represent sensor measurements. The inputs are comprised of w_l and w_m for feedback from the dynamic pressure perturbation, w_y and w_α as feedback from the uncertainty operators, and β as the position of the control surface.

The block diagram of the open-loop state-space model, P, and the feedback operator describing the repeated perturbation to dynamic pressure, $\delta_{\overline{q}}$, and uncertainty parameters, δ_{c_y} and δ_{k_α}, is shown in Figure 10.4.

The state-space model of (10.25) is formulated by including uncertainty in the equations of motion and then deriving corresponding state equations. This procedure is useful for considering uncertainties in specific parameters; however, it may be too cumbersome and lengthy for high-order systems that have many terms in the equations of motion.

An alternative method of deriving a state-space realization to model the robust aeroelastic system is to consider directly the state-space realization of the nominal aeroelastic system in (10.11).

Consider the update equation to compute $\ddot{y}$ from (10.9) and note which terms depend on uncertain parameters. Specifically, $a_{32} = a_{32_o} + a_{32_\alpha} k_\alpha$ is dependent on the pitch stiffness and $a_{33} = a_{33_y} + a_{33_y} c_y$ is dependent on the plunge damping. Thus, a robust model can be formulated by replacing the uncertain terms with their associated perturbed values. Separate the effects of the uncertainty parameters and replace with feedback signals.

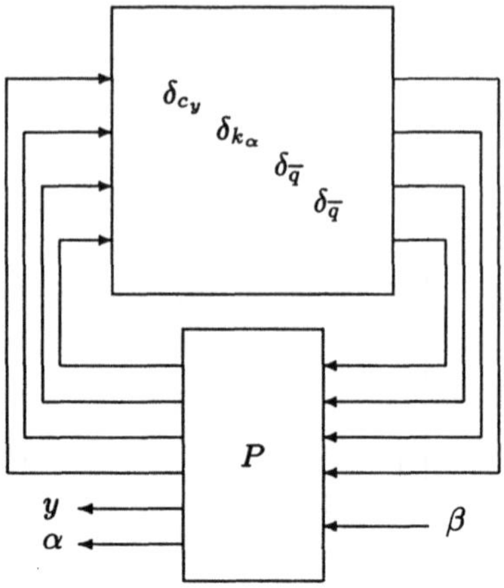

Fig. 10.4: Robust Aeroelastic Model of Pitch-Plunge System with Parameterization over Dynamic Pressure and Uncertainty Description

$$
\begin{aligned}
\ddot{y} &= a_{31}y + (a_{32_o} + a_{32_\alpha}k_\alpha)\,\alpha + \left(a_{33_o} + a_{33_y}c_y\right)\dot{y} + a_{34}\dot{\alpha} \\
&\quad + b_{3l}w_l + b_{3m}w_m + b_{3\beta}\beta \\
&= a_{31}y + (a_{32_o} + a_{32_\alpha}\left(k_{\alpha_o} + W_{k_\alpha}\delta_{k_\alpha}\right))\,\alpha \\
&\quad + \left(a_{33_o} + a_{33_y}\left(c_{y_o} + W_{c_y}\delta_{c_y}\right)\right)\dot{y} + a_{34}\dot{\alpha} \\
&\quad + b_{3l}w_l + b_{3m}w_m + b_{3\beta}\beta \\
&= a_{31}y + (a_{32_o} + a_{32_\alpha}k_{\alpha_o})\,\alpha + \left(a_{33_o} + a_{33_y}c_{y_o}\right)\dot{y} + a_{34}\dot{\alpha} \\
&\quad + b_{3l}w_l + b_{3m}w_m + a_{33_y}W_{c_y}\delta_{c_y}\dot{y} + a_{32_\alpha}W_{k_\alpha}\delta_{k_\alpha}\alpha + b_{3\beta}\beta \\
&= a_{31}y + (a_{32_o} + a_{32_\alpha}k_{\alpha_o})\,\alpha + \left(a_{33_o} + a_{33_y}c_{y_o}\right)\dot{y} + a_{34}\dot{\alpha} \\
&\quad + b_{3l}w_l + b_{3m}w_m + b_{3y}\delta_{c_y}W_{c_y}\dot{y} + b_{3\alpha}\delta_{k_\alpha}W_{k_\alpha}\alpha + b_{3\beta}\beta \\
&= a_{31}y + (a_{32_o} + a_{32_\alpha}k_{\alpha_o})\,\alpha + \left(a_{33_o} + a_{33_y}c_{y_o}\right)\dot{y} + a_{34}\dot{\alpha} \\
&\quad + b_{3l}w_l + b_{3m}w_m + b_{3y}\delta_{c_y}z_y + b_{3\alpha}\delta_{k_\alpha}z_\alpha + b_{3\beta}\beta \\
&= a_{31}y + (a_{32_o} + a_{32_\alpha}k_{\alpha_o})\,\alpha + \left(a_{33_o} + a_{33_y}c_{y_o}\right)\dot{y} + a_{34}\dot{\alpha} \\
&\quad + b_{3l}w_l + b_{3m}w_m + b_{3y}w_y + b_{3\alpha}w_\alpha + b_{3\beta}\beta
\end{aligned}
\tag{10.30}
$$

The explicit dependence of $\ddot{y}$ on the uncertainty parameters has been replaced with dependence on additional input and output signals in (10.30). The parametric uncertainty on plunge damping is now included with the system by the feedback relationship $w_y = \delta_{c_y}z_y$. Similarly, the parametric uncertainty on pitch stiffness is now included with the system by the feedback relationship $w_\alpha = \delta_{k_\alpha}z_\alpha$. The output signals used in these feedback relationships are easily seen from the derivation to be weighted pitch angle and plunge rate values.

$$z_y = W_{c_y}\dot{y} \tag{10.31}$$
$$z_\alpha = W_{k_\alpha}\alpha \tag{10.32}$$

The similarity of this derivation to the original derivation based on the equations of motion is immediately clear by comparing the formulations for z_y given in (10.15) and (10.31). These equations are identical and indicate the separate derivations produce identical formulations for the model.

Also, the coefficients that scale the states and input signals such as w_l, w_m and β in the equation to compute $\ddot{y}$ for the nominal aeroelastic model are identical to the corresponding coefficients in the equation to compute $\ddot{y}$ for the robust aeroelastic model. This agrees with the analysis of (10.19) that noted the same similarity of the coefficients for the model derived by including uncertainty in the original equations of motion. Thus, the robust aeroelastic model is similar for each derivation.

A similar derivation is performed on the equation to compute $\ddot{\alpha}$ in the equation for the nominal aeroelastic system given in (10.10).

$$\begin{aligned}
\ddot{\alpha} &= a_{41}y + (a_{42_o} + a_{42_\alpha}k_\alpha)\,\alpha + \left(a_{43_o} + a_{43_y}c_y\right)\dot{y} + a_{44}\dot{\alpha} \\
&\qquad + b_{4l}w_l + b_{4m}w_m + b_{4\beta}\beta \\
&= a_{41}y + \left(a_{42_o} + a_{42_\alpha}\left(k_{\alpha_o} + W_{k_\alpha}\delta_{k_\alpha}\right)\right)\alpha \\
&\qquad + \left(a_{43_o} + a_{43_y}\left(c_{y_o} + W_{c_y}\delta_{c_y}\right)\right)\dot{y} + a_{44}\dot{\alpha} \\
&\qquad + b_{4l}w_l + b_{4m}w_m + b_{4\beta}\beta \\
&= a_{41}y + \left(a_{42_o} + a_{42_\alpha}k_{\alpha_o}\right)\alpha + \left(a_{43_o} + a_{43_y}c_{y_o}\right)\dot{y} + a_{44}\dot{\alpha} \\
&\qquad + b_{4l}w_l + b_{4m}w_m + a_{43_y}W_{c_y}\delta_{c_y}\dot{y} + a_{42_\alpha}W_{k_\alpha}\delta_{k_\alpha}\alpha + b_{4\beta}\beta \\
&= a_{41}y + \left(a_{42_o} + a_{42_\alpha}k_{\alpha_o}\right)\alpha + \left(a_{43_o} + a_{43_y}c_{y_o}\right)\dot{y} + a_{44}\dot{\alpha} \\
&\qquad b_{4l}w_l + b_{4m}w_m + b_{4y}\delta_{c_y}W_{c_y}\dot{y} + b_{4\alpha}\delta_{k_\alpha}W_{k_\alpha}\alpha + b_{4\beta}\beta \\
&= a_{41}y + \left(a_{42_o} + a_{42_\alpha}k_{\alpha_o}\right)\alpha + \left(a_{43_o} + a_{43_y}c_{y_o}\right)\dot{y} + a_{44}\dot{\alpha} \\
&\qquad b_{4l}w_l + b_{4m}w_m + b_{4y}\delta_{c_y}z_y + b_{4\alpha}\delta_{k_\alpha}z_\alpha + b_{4\beta}\beta \\
&= a_{41}y + \left(a_{42_o} + a_{42_\alpha}k_{\alpha_o}\right)\alpha + \left(a_{43_o} + a_{43_y}c_{y_o}\right)\dot{y} + a_{44}\dot{\alpha} \\
&\qquad b_{4l}w_l + b_{4m}w_m + b_{4y}w_y + b_{4\alpha}w_\alpha + b_{4\beta}\beta
\end{aligned} \tag{10.33}$$

The uncertainty parameters have again been eliminated from the equation and replaced by signals that are related to the uncertainty operator by a feedback relationship. The derivation for $\ddot{\alpha}$ was performed independently of the derivation for $\ddot{y}$; however, the feedback signals to relate the uncertainty operators are identical. Thus, the uncertainty in plunge damping enters by the relationship $w_y = \delta_{c_y}z_y$ where z_y is given in (10.31). Similarly, the uncertainty in pitch stiffness enters by the relationship $w_\alpha = \delta_{k_\alpha}z_\alpha$ where z_α is given in (10.32).

The equation for $\ddot{y}$ given by (10.19) is identical to the equation for $\ddot{y}$ given by (10.30). Likewise, the equations for $\ddot{\alpha}$ given by (10.22) and (10.33) are matched. This equality indicates the state-space realization of the robust aeroelastic model that results from including parametric uncertainty directly in the equations of motion is identical to the model that results from including parametric uncertainty in the state-space realization of the nominal aeroelastic model. Thus, this example demonstrates that there are several methods to include uncertainty in an aeroelastic model.

The robust aeroelastic model may not be affected by the choice of derivation used to formulate the state-space matrices; however, there are certain advantages to deriving the robust model by including uncertainty in the state-space matrices of the nominal aeroelastic model. Firstly, this method requires much less algebra than the derivation that results by including uncertainty in the equations of motion. Also, this method allows uncertainty to be easily included in modal parameters, such as natural frequency and damping, that appear in state matrices but are not necessarily parameters in the equations of motion.

Alternatively, the derivation based on including uncertainty directly in the equations of motion also presents certain advantages that may be important for some applications. One advantage is the ability to directly include uncertainty in a specific parameter and ensure this uncertainty does not affect any other parameters. Another advantage is the ability to gain insight about which state-space elements are actually by a parameter and the amount of variation that results in modal elements as a result of uncertainty in a physical parameter. Of course, these issues are not very important for the low-order model of the pitch-plunge, but they may be relevant when considering high-order aircraft models.

10.4 Robust Aeroservoelastic Model

The procedure to analyze aeroservoelastic stability margins using μ can also be demonstrated with the pitch-plunge system. The position of the flap can be commanded by a feedback control system and introduce interactions that are not present in the open-loop aeroelastic model.

The equations of motion for this system are of relatively low order so the issue of control design may not seem too important; however, there are several challenges aspects of control design that have been associated with this testbed. In particular, nonlinear and adaptive strategies have been extensively studied for vibration attenuation and flutter margin augmentation [16, 67, 68, 69, 143].

A simple controller, K, is adopted for this example to command the flap based on feedback measurements. This controller is certainty not optimal for either vibration attenuation or flutter margin augmentation; however, it is sufficient to demonstrate the concept of robust aeroservoelastic stability.

$$\beta = K\alpha = -\alpha \tag{10.34}$$

The feedback matrix, K, is used as the only nontrivial element of the control system. This situation is not realistic for a complex system because actuators and sensors have dynamics that must be considered; however, this controller is sufficient for the tutorial nature of this example. Thus, the actuators and sensors are modeled as all-pass operators with unity gain over all frequency.

Also, the uncertainty description associated with the closed-loop model will be identical to the uncertainty description associated with the open-loop model. It is assumed there are no modeling errors in the control system because of the assumptions of ideal actuators and sensors. Again, a realistic model would have actuators and associated actuator uncertainties but these elements are not necessary to demonstrate the basic concepts of the μ method.

A disturbance signal is included with the closed-loop model to represent the effects of external inputs from possible wind gusts or excitation systems. This disturbance is modeled as acting through the actuator system and so the flap position of the model is actually the commanded flap position and the disturbance value.

Figure 10.5 presents the aeroservoelastic model that contains a feedback controller and an open-loop model with an associated uncertainty description.

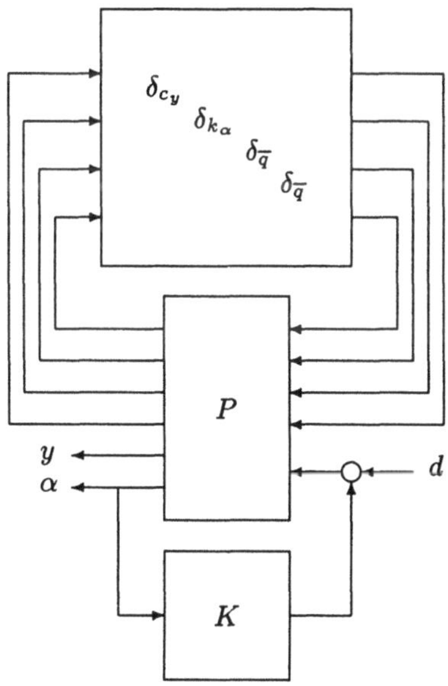

Fig. 10.5: Robust Aeroservoelastic Model of Pitch-Plunge System and Controller with Parameterization over Dynamic Pressure and Uncertainty Description

The robust aeroservoelastic model can be written by combining the open-loop equations of (10.25) and the controller feedback of (10.34).

$$\begin{bmatrix} \dot{y} \\ \dot{\alpha} \\ \ddot{y} \\ \ddot{\alpha} \\ \hline z_l \\ z_m \\ z_y \\ z_\alpha \\ y \\ \alpha \end{bmatrix} = \begin{bmatrix} A & B \\ \hline C & D \end{bmatrix} \begin{bmatrix} y \\ \alpha \\ \dot{y} \\ \dot{\alpha} \\ \hline w_l \\ w_m \\ w_y \\ w_\alpha \\ d \end{bmatrix} \tag{10.35}$$

The values of the state-space matrices are computed using the parameter values of Table 10.1 and are given in (10.36) through (10.39).

$$A = \begin{bmatrix} 0 & 0 & 1 & 0 \\ 0 & 0 & 0 & 1 \\ -291.112 & 7.304 & -3.390 & -0.063 \\ 1846.888 & -261.586 & 20.197 & -0.348 \end{bmatrix} \tag{10.36}$$

$$B = \begin{bmatrix} 0 & 0 & 0 & 0 & 0 \\ 0 & 0 & 0 & 0 & 0 \\ -0.102 & -0.649 & -0.102 & 0.649 & -0.085 \\ 0.649 & 19.504 & 0.649 & -19.504 & 2.071 \end{bmatrix} \tag{10.37}$$

$$C = \begin{bmatrix} 0 & 0.789 & 0.283 & 0.042 \\ 0 & -0.475 & -0.004 & -0.001 \\ 0 & 0 & 5 & 0 \\ 0 & 0.35 & 0 & 0 \\ 1 & 0 & 0 & 0 \\ 0 & 1 & 0 & 0 \end{bmatrix} \tag{10.38}$$

$$D = \begin{bmatrix} 0 & 0 & 0 & 0 & 0.009 \\ 0 & 0 & 0 & 0 & 0.004 \\ 0 & 0 & 0 & 0 & 0 \\ 0 & 0 & 0 & 0 & 0 \\ 0 & 0 & 0 & 0 & 0 \\ 0 & 0 & 0 & 0 & 0 \end{bmatrix} \tag{10.39}$$

The feedback controller is a relatively simple matrix that does not destabilize the system but it does alter the dynamics as evidenced by the difference between the closed-loop state matrix in (10.35) and the open-loop state matrix in (10.25). These differences result in a change in modal properties between the aeroelastic and aeroservoelastic models. The natural frequencies and dampings of the aeroservoelastic modes are given in Table 10.3.

Table 10.3: Modal Parameters for the Nominal Aeroservoelastic Pitch-Plunge Model

frequency (rad/s)	damping
12.646	.061
19.795	.055

The natural frequencies of the aeroservoelastic model in Table 10.3 are greater than the natural frequencies of the aeroelastic modes in Table 10.2 as a result of the feedback controller. Also, the damping values have changed such that one mode has increased damping but the other mode actually has less damping because of the controller.

These modal changes can be evidenced by computing the transfer functions from the disturbance input to the plunge measurement for each model. Figure 10.6 shows that peaks associated with modes for the aeroservoelastic model occur at noticeably higher frequencies than the peaks for the aeroelastic model. Also, the peak magnitude is greater near the natural frequency of the high-frequency mode and indicates the decrease in modal damping.

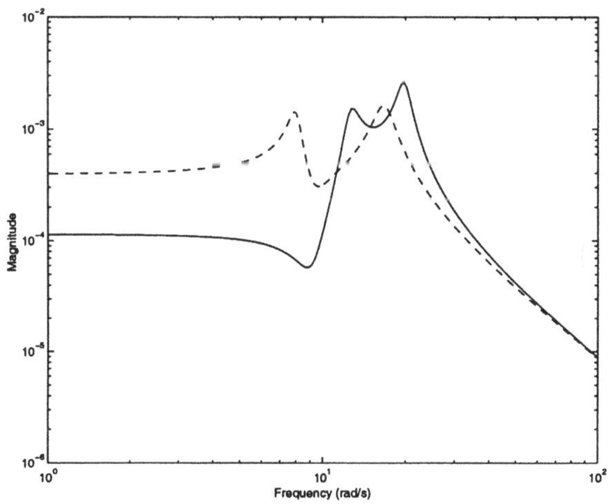

Fig. 10.6: Magnitude of Transfer Function from Noise to Plunge Measurement for Aeroelastic Model (– – –) and Aeroservoelastic Model (—)

Figure 10.7 shows the transfer functions for each model that relates the disturbance to the pitch measurement. This plot shows essentially the same information as Figure 10.6. The only noticeable difference is that the pitch measurement shows more clearly that the low-frequency mode has increased damping for the aeroservoelastic model as compared to the aeroelastic model.

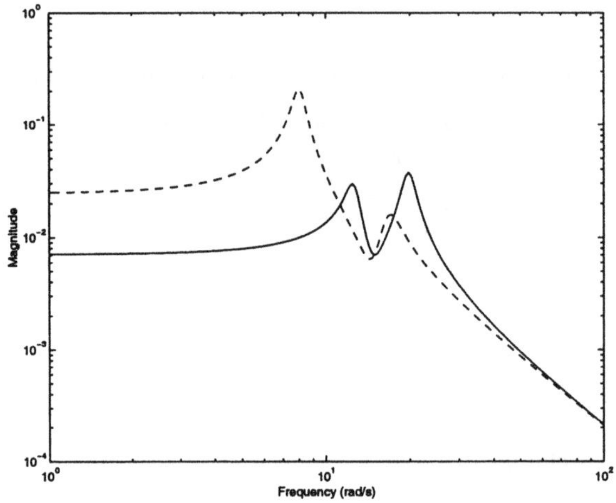

Fig. 10.7: Magnitude of Transfer Function from Noise to Pitch Measurement for Aeroelastic Model (– – –) and Aeroservoelastic Model (—)

Time responses are simulated to demonstrate how the open-loop and closed-loop system behave. These responses are computed as free decay motion after the states of the model have been set to the values given in Table 10.4.

Table 10.4: Initial Conditions for Time Simulation

state	initial value
y	$0\ m$
α	$.1\ rad$
$\dot{y}$	$0\ m/s$
$\dot{\alpha}$	$0\ rad/s$

Figure 10.8 demonstrates the plunge motion in response to the initial condition is greater for the aeroservoelastic model as compared to the aeroelastic model. This increase in amplitude corresponds to the decrease in modal damping indicated in Figure 10.6.

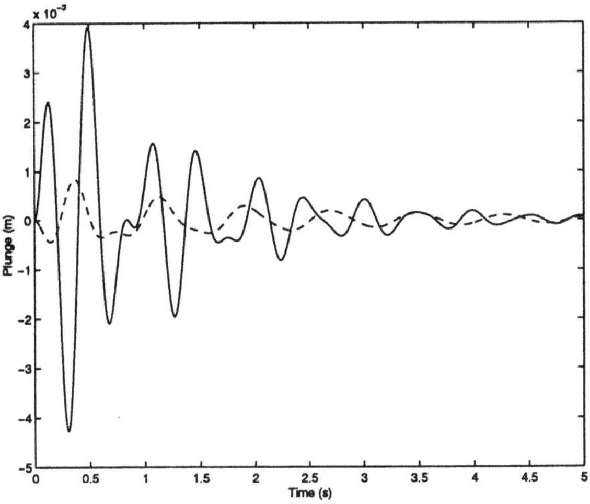

Fig. 10.8: Time Response of Plunge Measurement from Initial Condition for Aeroelastic Model (— — —) and Aeroservoelastic Model (—)

The associated pitch response is shown in Figure 10.9 and it demonstrates the increase in damping that is shown in Figure 10.7 as result of including the control system.

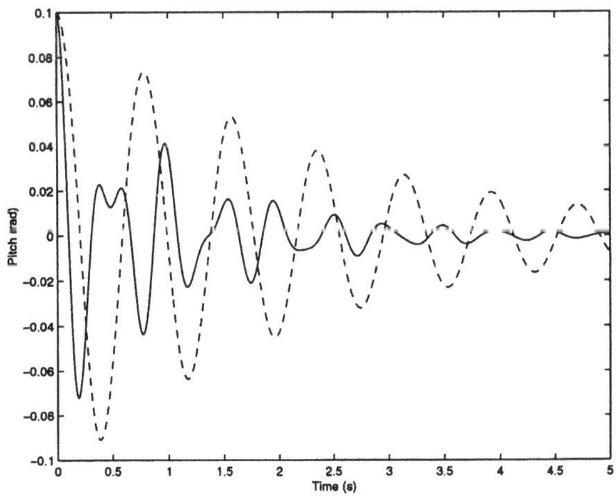

Fig. 10.9: Time Response of Pitch Measurement from Initial Condition for Aeroelastic Model (— — —) and Aeroservoelastic Model (—)

10.5 Aeroelastic Stability Margins

A simple way to estimate nominal flutter margins is to compute an aeroelastic model from (10.1) at different values of dynamic pressure and analyze the modal damping in a fashion similar to $V - g$ analysis. Consider the plots of modal damping in Figure 10.10 and natural frequency in Figure 10.11 shown for a range of dynamic pressures for the nominal pitch-plunge model.

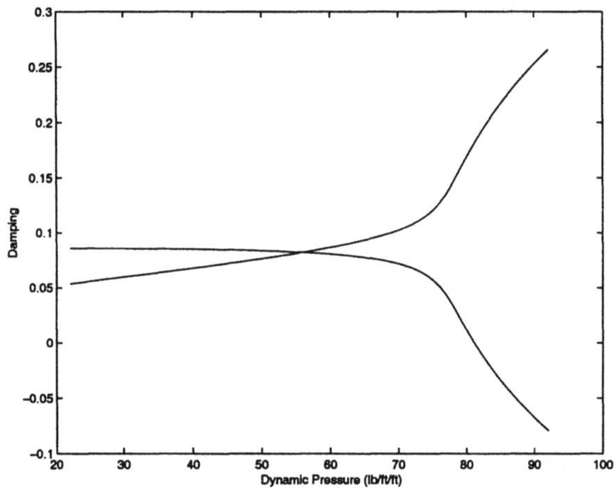

Fig. 10.10: Modal Dampings of the Nominal Open-Loop Model

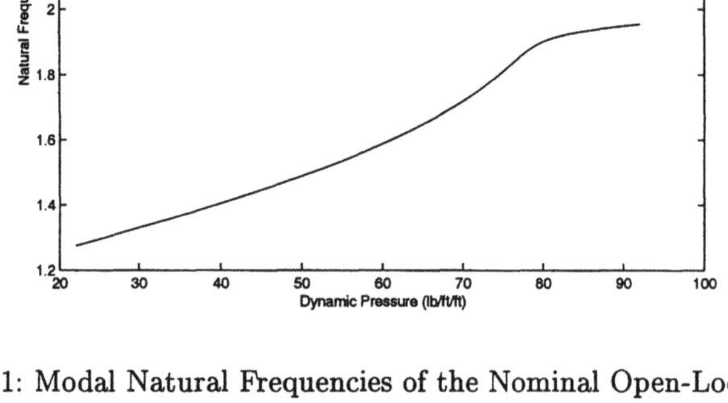

Fig. 10.11: Modal Natural Frequencies of the Nominal Open-Loop Model

The nominal flutter margin corresponds to the smallest value of dynamic pressure at which the model becomes unstable. This value represents the critical flutter pressure and is indicated by the dynamic pressure at which the damping for at least one mode becomes negative. Figure 10.10 clearly shows the nominal flutter pressure is $\overline{q} = 81\ lb/ft^2$.

Figure 10.11 does not directly indicate the flutter margin; however, it is still of interest for analyzing flutter behaviors. This plot indicates that the natural frequencies coalese as the dynamic pressure increases. Thus, the pitch-plunge model experiences a classical flutter mechanism that results from the structural modes coupling.

Nominal flutter margins are also computed using the μ method on the aeroelastic model in (10.11). These margins are computed by varying the perturbation to dynamic pressure using Algorithm 8.1.2 and finding the smallest destabilizing perturbation. The critical flutter pressure computed using μ of $\overline{q} = 81\ lb/ft^2$ is nearly identical to Γ_{V-g} and indicates the μ method can be used instead of classical methods like V-g analysis.

Robust flutter margins are computed by considering the aeroelastic model with uncertainties given in (10.25). These margins indicate the smallest perturbation to dynamic pressure for which the aeroelastic model is not robustly stable to the modeled uncertainties in plunge damping and pitch stiffness. The critical flutter pressure for the robust model is $\overline{q} = 76\ lb/ft^2$.

Flutter margins are computed as the difference between the operating dynamic pressure and the critical pressure associated with a flutter instability. These margins and the frequencies of the modes associated with the critical flutter mechanism are given in Table 10.5..

Table 10.5: Flutter Margins and Frequencies of Critical Modes

method	Γ	ω
V-g	$59\ lb/ft^2$	$2.00\ Hz$
μ_{nom}	$59\ lb/ft^2$	$2.06\ Hz$
μ_{rob}	$54\ lb/ft^2$	$2.12\ Hz$

The robust flutter margins are less than the nominal flutter margins and demonstrate the effects of including uncertainty in the analysis. In this case, the robust margin suggests that the true flutter margin may be $4\ lb/ft^2$ less than is indicated by the nominal aeroelastic model. The nominal and robust flutter margins are not significantly different for the pitch-plunge model; however, they do demonstrate that even a small amount of uncertainty can affect the flutter analysis.

Also, the frequencies are nearly identical for the modes associated with the nominal and robust flutter margins so the critical flutter mechanism is essentially the same. Thus, the uncertainty affects the dynamic pressure at which flutter occurs but it does not signficantly affect the critical mode.

10.6 Aeroservoelastic Stability Margins

Nominal aeroservoelastic stability margins can be estimated by applying the simple approach based on $V - g$ analysis that computes models at a series of dynamic pressures. The resulting values of modal damping are shown in Figure 10.12 and corresponding natural frequencies are shown in Figure 10.13 for a range of conditions.

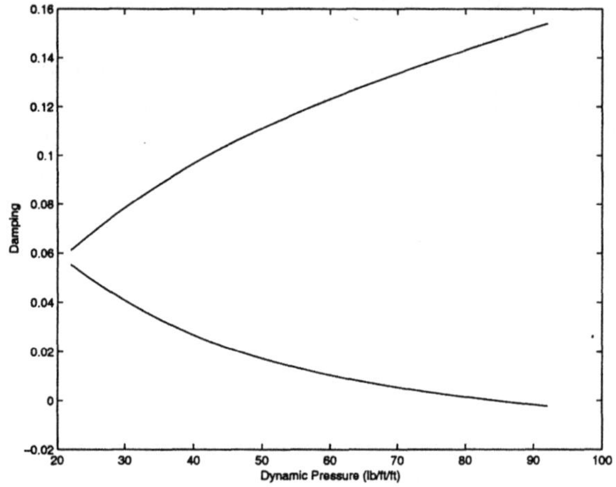

Fig. 10.12: Modal Dampings of the Nominal Closed-Loop Model

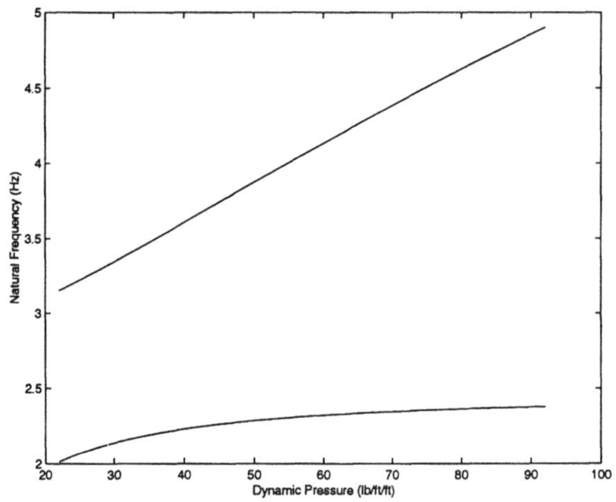

Fig. 10.13: Modal Natural Frequencies of the Nominal Closed-Loop Model

The evolution of modal properties for the aeroservoelastic model are considerably different than the corresponding evolution computed for flutter analysis. Notably, the natural frequencies do not coalese as dynamic pressure increases so the modes are not coupling. Another difference is that the critical mode has a slowly decreasing damping value instead of a sudden decrease near the critical flutter pressure. Thus, the aeroservoelastic dynamics do not experience a classical flutter mechanism because of the feedback controller.

An estimate of the critical pressure at which an aeroservoelastic instability is encountered is shown by the condition at which a damping value in Figure 10.12 becomes negative. This pressure is approximately $\overline{q}_{V-g} = 84\ lb/ft^2$ and the mode occurs at approximately $\omega_{V-g} = 4.7\ Hz$.

Nominal aeroservoelastic stability margins are computed by applying Algorithm 8.1.2 to the closed-loop model. The μ method indicates that an instability is encountered at $\overline{q}_{nom} = 84.4\ lb/ft^2$ for a mode with frequency at $\omega_{nom} = 4.7\ Hz$.

The μ method is also used to compute robust aeroservoelastic stability margins by analyzing the closed-loop model with uncertainties given in (10.35). The critical pressure at which the dynamics are not robust stable to the uncertainty description is $\overline{q}_{rob} = 50.6\ lb/ft^2$. The associated critical mode has a natural frequency of $\omega_{rob} = 3.9\ Hz$.

The stability margins which relate the difference between the operating pressure and the critical pressure associated with an aeroelastic instability are given in Table 10.6.

Table 10.6: Aeroservoelastic Stability Margins and Frequencies of Critical Modes

method	Γ	ω
V-g	$59\ lb/ft^2$	$4.7\ Hz$
μ_{nom}	$62\ lb/ft^2$	$4.7\ Hz$
μ_{rob}	$28\ lb/ft^2$	$3.9\ Hz$

The value of Γ_{rob} is considerably lower than Γ_{nom} and indicates that the aeroservoelastic dynamics are highly sensitive to the modeling uncertainties shown in Figure 10.5. Nominally, the feedback controller is able to extend the stability margin beyond the flutter margin; however, the effect of the uncertainty is considerably greater on the aeroservoelastic stability margin than the flutter margin. Thus, this example demonstrates the importance of considering uncertainty when analyzing feedback controllers for aeroelastic systems. Traditional analysis indicates the controller is beneficial for expanding a flight envelope but the μ method demonstrates that this result may be misleading and the aeroservoelastic model may actually be closer to an instability than the open-loop model.

The reason for the high sensitivity to uncertainty is easily seen for this simply example by considering Figure 10.12. This plot shows the modal damping decreases to a value very near 0 as the dynamic pressure increases to about 50 lb/ft^2. Thus, the variations admitted by modeling uncertainty present enough of a range in modal properties that negative dampings are included in the set of robust plant models.

Also, the frequency associated with the critical aeroservoelastic instability changes by approximately 20% when uncertainty is included in the analysis. This may see like a fairly large change and possibly indicates a change in the critical mode; however, analysis of Figure 10.13 indicates this level of change in frequency is expected. The critical mode shows a large change in natural frequency as the dynamic pressure is increased and so evaluating the modal properties at a somewhat smaller dynamic pressure should result in a large decrease in frequency.

A.10 Computer Code

The analysis of the pitch-plunge system is presented as an instructive tutorial that demonstrates the μ method as applied to an aeroservoelastic system. Towards that end, it may be useful to have a simulation that enables the results presented here to be reproduced. Also, it may be useful to consider alterations to the simulation to easily investigate how the results are altered by changes such as different model parameters and different uncertainty descriptions.

The procedures used to compute the robust and nominal stability margins for the pitch-plunge system are essentially given by Algorithm 8.1.4 and Algorithm 8.2.2; however, it may be difficult to convert these algorithms into executable computer code. The difficulty lies in writing routines to efficiently and accurately compute the upper bound for μ. Thus, it is prudent to consider using a commercial software package.

All the results presented in this text were generated using MATLAB©[1]. In particular, the routines for computing μ were provided by the $\mu - Tools$ toolbox and are very straightforward to use.

This section presents the computer code for MATLAB that can be used to reproduce the analysis of the pitch-plunge system. The program should generate several plots that match the figures from this chapter and also print the information about the stability margins. This information relates the dynamic pressure at the critical instability and the frequency of the mode that becomes unstable.

```
Nominal Flutter at qbar = 81.1473 and w = 2.06316
Robust  Flutter at qbar = 76.6508 and w = 2.11521
Nominal ASE     at qbar = 84.4249 and w = 4.72836
Robust  ASE     at qbar = 50.6451 and w = 3.89066
```

This program can be run exactly as typed. There are no input data files it needs to read and there are no output data files that are created.

[1] MATLAB is a registered trademark of The MathWorks, Inc.

```
%---------------------------------------------------------------
% Define System Parameters
%---------------------------------------------------------------


U = 6;                              % airspeed
e=-0.6;                             % elastic axis

b=.135;
p=1.225;                            % density of air
m=12.387;                           % mass
I=0.065;                            % moment of inertia
ky=2844.4;                          % plunge stiffness
ka=3.525;                           % pitch stiffness
cy=27.43;                           % plunge damping
ca=.036;                            % pitch damping
cla=6.28;                           % lift coefficient
clb=3.358;                          % lift coefficient
cma=-.635;                          % moment coefficient
cmb=12.387;                         % moment coefficient
x=.2466;                            % parameter for span
E = .5-e;                           % parameter for axis

q = .5*p*U*U;                       % nominal dynamic pressure

Wy = 5.0;                           % weighting on zy
Wk = 0.35;                          % weighting on zk

%---------------------------------------------------------------
% Define Elements of State-Space Matrices
%---------------------------------------------------------------

zla = 2*b*cla;                      % coefficient of zl
zlyy = 2*b*cla/U;                   % coefficient of zl
zlaa = 2*b*cla*E*b/U;               % coefficient of zl
zlb = 2*b*clb;                      % coefficient of zl

zma = 2*b*b*cma;                    % coefficient of zm
zmyy = 2*b*b*cma/U;                 % coefficient of zm
zmaa = 2*b*b*cma*E*b/U;             % coefficient of zm
zmb = 2*b*b*cmb;                    % coefficient of zm
```

```
Y = 1/(I*m-m*m*x*x*b*b);          % scaling for plunge state
A = 1/(I-m*x*x*b*b);              % scaling for pitch state

a31 = Y*(-I*ky);
a32a = Y*(m*x*b);
a32o = Y*(-q*2*(I*b*cla + m*x*b^3*cma));
a41 = A*(x*b*ky);
a42a = A*(-1);
a42o = A*(q*2*(b*b*cma + x*b*b*cla));
a33y = Y*(-I);
a33o = Y*(-q*2*(I*b*cla + m*x*b^3*cma)/U);
a34 = Y*(m*x*b*ca - q*2*(I*b*cla+m*x*b^3*cma)*E*b/U);
a43y = A*(x*b);
a43o = A*(q*2*(b*b*cma + x*b*b*cla)/U);
a44 = A*(-ca + q*2*(b*b*cma+x*b*b*cla)*E*b/U);

b3l = Y*(-I);
b3m = Y*(-m*x*b);
b3y = b3l;
b3a = -b3m;
b3b = Y*(-q*2*(I*b*clb+m*x*b^3*cmb));
b4l = A*(x*b);
b4m = A*(1);
b4y = b4l;
b4a = -b4m;
b4b = A*(q*2*(b*b*cmb+x*b*b*clb));

A = [  0        0              1           0;
       0        0              0           1;
       a31  a32o+a32a*ka  a33o+a33y*cy  a34;
       a41  a42o+a42a*ka  a43o+a43y*cy  a44];

B = [0   0   0   0   0;
     0   0   0   0   0;
     b3l b3m b3y b3a b3b;
     b4l b4m b4y b4a b4b];

C = [0 zla zlyy zlaa;
     0 zma zmyy zmaa;
     0  0   Wy    0;
     0  Wk  0     0;
     1  0   0     0;
     0  1   0     0];
```

```
D = [0 0 0 0 zlb;
     0 0 0 0 zmb;
     0 0 0 0 0;
     0 0 0 0 0;
     0 0 0 0 0;
     0 0 0 0 0];

%-----------------------------------------------------------
% Make Open-Loop System
%-----------------------------------------------------------

P = pck(A,B,C,D);                % name open-loop model

%-----------------------------------------------------------
% Make Closed-Loop System
%-----------------------------------------------------------

K = [0 -1];                      % controller

systemnames = 'P K';             % closed-loop elements
inputvar = '[w(4);n]';           % inputs to system
outputvar = '[P]';               % outputs of system
input_to_P = '[w;K+n]';          % inputs to plant
input_to_K = '[P(5:6)]';         % feedback plunge,pitch
sysoutname = 'G';                % name closed-loop model
sysic                            % make closed-loop model

%-----------------------------------------------------------
% Plot Time Responses
%-----------------------------------------------------------

xinit = [0;.1;0;0];              % initial condition
in = zeros(5,1);                 % input signal
olp = trsp(P,in,5,.01,xinit);    % open-loop response
clp = trsp(G,in,5,.01,xinit);    % closed-loop response

olpy = sel(olp,5,1);             % open-loop plunge
olpa = sel(olp,6,1);             % open-loop pitch

clpy = sel(clp,5,1);             % closed-loop plunge
clpa = sel(clp,6,1);             % closed-loop pitch
```

```
figure(1)                          % open figure window
subplot(211)                       % top half
vplot(olpy,'r',clpy,'b');          % plot plunge
ylabel('Plunge (m)');              % axis label
xlabel('Time (s)');                % axis label
subplot(212)                       % bottom half
vplot(olpa,'r',clpa,'b');          % plot pitch
ylabel('Pitch (rad)');             % axis label
xlabel('Time (s)');                % axis label

%-----------------------------------------------------------
% Plot Transfer Functions from Noise to Measurements
%-----------------------------------------------------------
w = logspace(0,2,200);             % frequency range
gp = frsp(P,w);                    % transfer function
gg = frsp(G,w);                    % transfer function

olpny = sel(gp,5,5);               % open-loop plunge
olpna = sel(gp,6,5);               % open-loop pitch
clpny = sel(gg,5,5);               % closed-loop plunge
clpna = sel(gg,6,5);               % closed-loop pitch

figure(2)                          % open figure window
subplot(211)                       % top half
vplot('liv,lm',olpny,'r',clpny,'b');
ylabel('Magnitude');               % axis label
xlabel('Frequency (rad/s)');       % axis label
subplot(212)                       % bottom half
vplot('liv,lm',olpna,'r',clpna,'b');
ylabel('Magnitude');               % axis label
xlabel('Frequency (rad/s)');       % axis label

%-----------------------------------------------------------
% Plot Modal Properties of Open-Loop System
%-----------------------------------------------------------
qdel = 0:1:70;                             % perturbations

for i=1:length(qdel)                       % search range
    sys = starp(qdel(i)*eye(2),P);         % close qbar
    ee = spoles(sys)';                     % poles
    zvec(i,:) = -real(ee)./abs(ee);        % dampings
    wvec(i,:) = abs(ee)/6.28;              % frequencies
end
```

```
figure(3)                                   % open figure
subplot(211)                                % top half
plot(q+qdel,zvec,'g');                      % plot dampings
ylabel('Damping');                          % axis label
xlabel('Dynamic Pressure (lb/ft/ft)');      % axis label
subplot(212)                                % bottom half
plot(q+qdel,wvec,'g');                      % plot frequencies
ylabel('Frequency (Hz)');                   % axis label
xlabel('Dynamic Pressure (lb/ft/ft)');      % axis label

%-----------------------------------------------------------
% Plot Modal Properties of Closed-Loop System
%-----------------------------------------------------------
qdel = 0:1:70;                              % perturbations

for i=1:length(qdel)                        % search range
    sys = starp(qdel(i)*eye(2),G);          % close qbar
    ee = spoles(sys)';                      % poles
    zvec(i,:) = -real(ee)./abs(ee);         % dampings
    wvec(i,:) = abs(ee)/6.28;               % frequencies
end

figure(4)                                   % open figure
subplot(211)                                % top half
plot(q+qdel,zvec,'g');                      % plot dampings
ylabel('Damping');                          % axis label
xlabel('Dynamic Pressure (lb/ft/ft)');      % axis label
subplot(212)                                % bottom half
plot(q+qdel,wvec,'g');                      % plot frequencies
ylabel('Frequency (Hz)');                   % axis label
xlabel('Dynamic Pressure (lb/ft/ft)');      % axis label

%-----------------------------------------------------------
% Nominal Aeroelastic Stability Margin
%-----------------------------------------------------------

del_min = 0;                        % minimum perturbation
del_max = 300;                      % maximum perturbation
while del_max - del_min > .01       % set tolerance
    del = (del_max+del_min)/2;      % current perturbation
    sys = starp(del*eye(2),P);      % close qbar loop
    ee = spoles(sys);               % poles
    if max(real(ee)) > 0            % check for instability
```

```
        del_max = del;            % new upper bound
    else                          % check for stability
        del_min = del;            % new lower bound
    end
end

qbar_nom = q + del_max;           % flutter pressure

sys = starp(del_max*eye(2),P);  % unstable model
ee = spoles(sys);                % poles
[ereal,ind] = sort(real(ee));    % sort real component
pole = ee(ind(max(size(ind)))); % unstable pole
w_nom = abs(pole)/6.28;          % flutter frequency

fprintf('Nominal Flutter at qbar = %g and w = %g\n', ...
                qbar_nom,w_nom);

%-----------------------------------------------------------
% Robust Aeroelastic Stability Margin
%-----------------------------------------------------------
Pmu = sel(P,1:4,1:4);            % ignore sensors
blk_mu = [-2 0;-1 0;-1 0];       % block structure
mu_bnd = 1/del_max;              % upper bound for mu
mu_val = 1.1;                    % initial value for mu
w_mu = 0:.1:18;                  % frequency range
while mu_val > 1.05              % enforce mu <= 1
    WW=eye(2)/mu_bnd;            % scaling on dqbar
    PP=mmult(daug(WW,1,1),Pmu); % scale plant
    g = frsp(PP,w_mu);           % frequency response
    b = mu(g,blk_mu,'Uws');      % compute mu
    mu_val = pkvnorm(b,'inf');   % upper bound for mu
    mu_bnd = mu_bnd*mu_val;      % upper bound on flutter
end

qbar_rob = q + 1/mu_bnd;         % flutter pressure

sys = starp(eye(2)/mu_bnd,P);   % unstable model
ee = spoles(sys);                % poles
[ereal,ind] = sort(real(ee));    % sort real component
pole = ee(ind(max(size(ind)))); % unstable pole
w_rob = abs(pole)/6.28;          % flutter frequency

fprintf('Robust  Flutter at qbar = %g and w = %g\n', ...
                qbar_rob,w_rob);
```

```
%-------------------------------------------------------------
% Nominal Aeroservoelastic Stability Margin
%-------------------------------------------------------------

del_min = 0;                        % minimum perturbation
del_max = 300;                      % maximum perturbation
while del_max - del_min > .01       % set tolerance
    del = (del_max+del_min)/2;      % current perturbation
    sys = starp(del*eye(2),G);      % close qbar loop
    ee = spoles(sys);               % poles
    if max(real(ee)) > 0            % check for instability
        del_max = del;              % new upper bound
    else                            % check for stability
        del_min = del;              % new lower bound
    end
end

qbar_nom = q + del_max;             % flutter pressure

sys = starp(del_max*eye(2),G);      % unstable model
ee = spoles(sys);                   % poles
[ereal,ind] = sort(real(ee));       % sort real component
pole = ee(ind(max(size(ind))));     % unstable pole
w_nom = abs(pole)/6.28;             % flutter frequency

fprintf('Nominal ASE     at qbar = %g and w = %g\n', ...
               qbar_nom,w_nom);

%-------------------------------------------------------------
% Robust Aeroservoelastic Stability Margin
%-------------------------------------------------------------

Gmu = sel(G,1:4,1:4);               % ignore sensors
blk_mu = [-2 0;-1 0;-1 0];          % block structure
mu_bnd = 1/del_max;                 % upper bound for mu
mu_val = 1.1;                       % initial value for mu
while mu_val > 1.05                 % enforce mu <= 1
    WW=eye(2)/mu_bnd;               % scaling on dqbar
    GG=mmult(daug(WW,1,1),Gmu);     % scale plant
    g = frsp(GG,w_mu);              % frequency response
    b = mu(g,blk_mu,'Uws');         % compute mu
    mu_val = pkvnorm(b,'inf');      % upper bound for mu
    mu_bnd = mu_bnd*mu_val;         % upper bound on flutter
end
```

```
qbar_rob = q + 1/mu_bnd;         % flutter pressure

sys = starp(eye(2)/mu_bnd,G);    % unstable model
ee = spoles(sys);                % poles
[ereal,ind] = sort(real(ee));    % sort real component
pole = ee(ind(max(size(ind)))); % unstable pole
w_rob = abs(pole)/6.28;          % flutter frequency

fprintf('Robust  ASE     at qbar = %g and w = %g\n', ...
                qbar_rob,w_rob);
```

CHAPTER 11

ROBUST FLUTTER MARGINS OF THE F / A-18 SRA

11.1 Flight Flutter Test

Robust flutter margins are computed for the F/A-18 Systems Research Aircraft, SRA, which is being flown at NASA Dryden Flight Research Center as Ship 845 [120]. The SRA is a standard two seat F/A-18 with production engines. This aircraft is being used as a testbed for flutter testing, advanced actuator concepts, smart structures, optical sensors and avionics systems [121]. The F/A-18 SRA is shown in flight over the Mojave desert near Edwards, CA in Figure 11.1.

Fig. 11.1: F/A-18 Systems Research Aircraft

Flutter testing was initiated on the SRA due to a major left wing structural modification to allow testing of several hydraulic and electromechanical aileron actuator concepts. The increased size and weight of these actuators required larger and heavier items to be installed replacing a fitting called a hinge-half which supports the aileron hinge, actuator, and a fairing. These structural modifications changed stiffness and damping properties along with adding about 20 *lb* to the wing. Dependency of the aileron aeroelastic behavior on actuator dynamics warranted the flutter tests.

A flutter analysis and associated flight flutter tests are performed on the F/A-18 SRA with the left wing structural modification. Over 30 flight tests were conducted in two sessions between September 1994 until February 1995 and between June 1995 until July 1995. A total of 260 data sets are measured from test points at various conditions throughout the flight envelope.

The flight flutter test program utilized a wingtip excitation system developed by Dynamic Engineering Incorporated (DEI) [110]. This DEI exciter is a modification of an excitation system used for F-16 XL flutter research [132]. The system consists of a wingtip exciter, an avionics box mounted in the instrumentation bay, and a cockpit controller.

Aerodynamic forces are generated by the wingtip exciter [81]. This exciter consists of a small fixed aerodynamic vane forward of a rotating slotted hollow cylinder. Rotating the cylinder varies the pressure distribution on the vane and results in a wingtip force changing at twice the cylinder rotation frequency. The magnitude of the resulting force is determined by the amount of opening in the slot. The F/A-18 aircraft with a left side wingtip exciter in the aft position is shown in Figure 11.2.

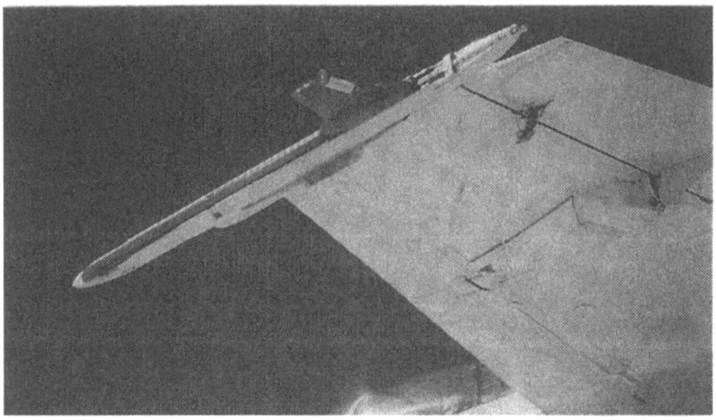

Fig. 11.2: Left Wing of the F/A-18 Systems Research Aircraft with Wingtip Excitation System

The cockpit controller commands a frequency range, duration and magnitude for the wingtip excitation signal. Frequency-varying excitation is generated by changing the cylinder rotation frequency with sine sweeps. Each wingtip exciter is commanded to act in-phase, 0 *degrees*, or out-of-phase, 180 *degrees*, with each other. Ideally, the in-phase data excites the symmetric modes of the aircraft and the out-of-phase data excites the antisymmetric modes although, in reality, non-ideal phasing of the exciters often caused a mixing of symmetric and antisymmetric modes to be excited.

Flight data sets are recorded by activating the exciter system at a given flight condition. The aircraft attempts to remain at the flight condition throughout the series of sine sweeps desired by the controller. The sine sweeps were restricted to within 3 *Hz* and 35 *Hz* with smaller ranges sometimes used to concentrate on a specific set of modal responses. Multiple sets of sweeps with varying parameters were tested.

The flight test matrix of excitation parameters and flight conditions for the F/A-18 SRA flutter program is given in Table 11.1.

Table 11.1: Parameters for the F/A-18 SRA Flight Flutter Test

Mach	0.54,0.65,0.70,0.80,0.85,0.90,0.95,1.05,1.2,1.4,1.6
Altitude (ft)	10k, 30k and 40k
Exciter configuration	Symmetric,antisymmetric and independent
Exciter position	Both forward, both aft and left-aft/right-forward
Sweep type	Linear and logarithmic
Sweep duration (sec)	15, 30 and 60
Sweep range (Hz)	3-35, 3-12, 3-25, 25-35, 35-3, 3-40
Sweep multiples	1,2 and 4 sweeps per maneuver

Accelerometers are available at several points on the aircraft to record modal responses. Each wing has a sensor on the aileron and at the forward and aft position on the wingtip. Additional accelerometers are located on each vertical tail and horizontal stabilator.

11.2 Flight Data Analysis

Aeroelastic flight data generated with the DEI exciter system is analyzed by generating transfer functions from the excitation force to the sensor measurements. These transfer functions are generated using standard Fourier transform algorithms. There are several inherent assumptions associated with Fourier analysis that are violated with the flight data. Particularly, the assumptions of time-invariant stationary data composed of sums of infinite sinusoids is not met by this transient response data [26, 80]. Alternative processing techniques, such as wavelet analysis, do not have these assumptions and so can accurately process data and potentially reduce the uncertainty required to validate the model [25]; however, standard Fourier routines provide a reasonable estimation of the data properties.

Figure 11.3 presents an example of dynamical variations that are observed during flight testing. Two transfer functions generated from different data sets clearly show some variation. The natural frequency and modal response is different as observed by these two data sets even though each test point utilized the same sweep procedure at identical flight conditions of Mach 0.8 and 30 *kft*. One possibility to account for such variations is the change in mass, up to 10%, between the heavyweight and lightweight conditions due to fuel consumption. Another possibility is the non-ideal phasing between the exciters causing excitation which is not purely symmetric.

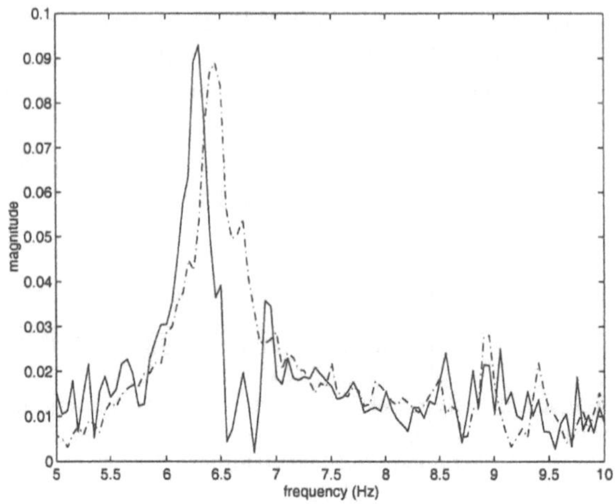

Fig. 11.3: Flight Data Transfer Functions from Symmetric Excitation to Left Wing Forward Accelerometer for Mach=.8 and 30000 feet Demonstrating Variation in Modal Frequency and Damping

The deviations between the modal response in Figure 11.3 are used to generate parametric uncertainty for the symmetric Wing 1st Bending mode. Similar deviations observed in modal responses are used to generate uncertainty for each mode. Uncertainty levels in modes which are not observed are assumed to be similar to those of the observed modes.

Figure 11.3 demonstrates how uncertainty is required to model variations between the analytical model and the aircraft dynamics which change during a flight test. Additional uncertainty is introduced due to improper operation of the DEI exciters. Several behaviors are observed due to these improper operations.

The most noticeable of these improper behaviors occurred as a function of dynamic pressure. The motors and rotating cylinder assemblies would bind at high dynamic pressures causing erratic phase relationships between the individual exciters on opposite wingtips. Figure 11.4 shows a representative plot of the phase difference in *degrees* between the left and right exciters operating in symmetric mode for a sweep taken at a test point with $\overline{q} = 356\ lb/ft^2$. This phase plot, which does not maintain the desired 0 *degrees* phase difference but is relatively constant and well behaved, is in direct contrast to the phase difference at a test point of $\overline{q} = 825\ lb/ft^2$ shown in Figure 11.5.

Another improper behavior is the nonrepeatibility observed with the data due to variations in transfer functions using different data sets recorded at identical flight conditions. An example of such a variation observed for frequency increasing and decreasing sweeps is shown in Figures 11.6 and 11.7.

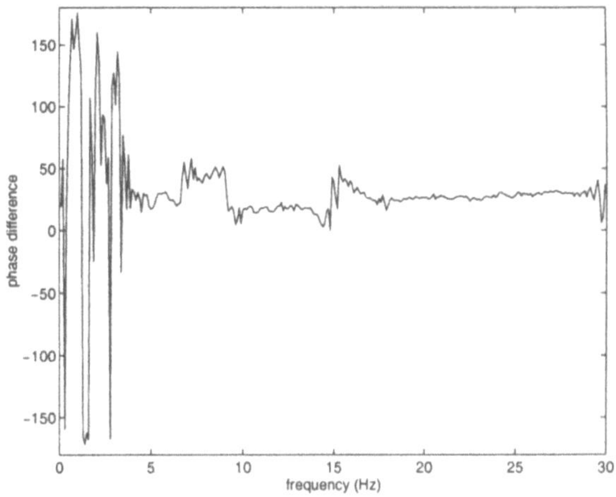

Fig. 11.4: Phase Difference in *degrees* between Left and Right Exciters for Sweep Range 3-30 Hz at Mach=.90 and 30000 feet at $\bar{q} = 356\ lb/ft^2$

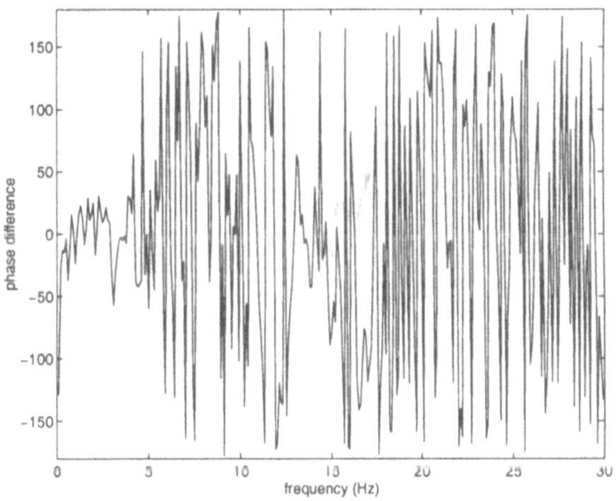

Fig. 11.5: Phase Difference in *degrees* between Left and Right Exciters for Sweep Range 3-30 Hz at Mach=.90 and 10000 feet at $\bar{q} = 825\ lb/ft^2$

The μ method is able to utilize the data despite these nonrepeatabilities which would adversely affect traditional system identification approaches.

The data analysis process is further degraded by a poor measurement of the excitation force used to generate the modal responses. The excitation force is not directly measured but rather a strain gauge measurement is used to approximate this force. The strain gauge records lateral shear strain at the exciter vane root. This measurement is assumed to be representative

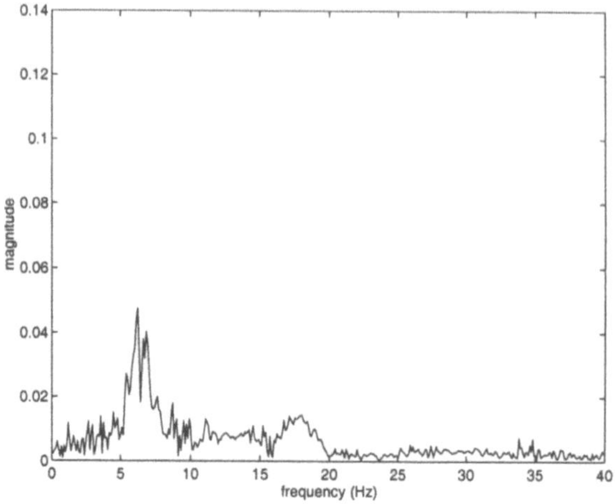

Fig. 11.6: Flight Data Transfer Function from Symmetric Excitation to Left Wing Forward Accelerometer for Mach=.8 and 30000 feet with Frequency Increasing Sweep from 3-35 Hz

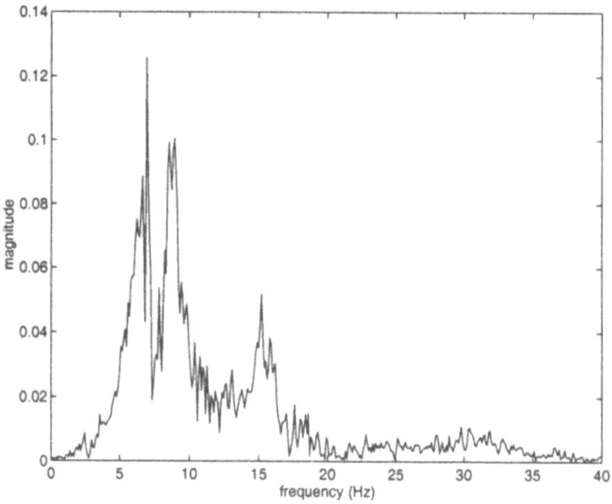

Fig. 11.7: Flight Data Transfer Function from Symmetric Excitation to Left Wing Forward Accelerometer for Mach=.8 and 30000 feet with Frequency Decreasing Sweep from 35-3 Hz

of the vertical shear and spanwise moment load at the wingtip rib. Time frequency analysis using wavelets clearly shows this strain gage signal has some harmonic resonances from the structural response [26].

The effect of the poor approximation to input force and the erratic behavior of the exciters is to reduce the quality of the flight data. Methods relying on system identification fail to accurately utilize the data to predict a flutter boundary [36]. The μ method is able to account for the data anomalies by including greater levels of uncertainty.

11.3 Analytical Model

A finite element model of the SRA is used to compute the aeroelastic dynamics and modal characteristics of the aircraft. The generalized equations of motion are used to derive a linear, finite-dimensional state-space model of the aircraft. This model contains 14 symmetric structural modes, 14 antisymmetric structural modes and 6 rigid-body dynamic modes. The control surfaces are not active and no control modes are included in the model. Natural frequencies of the elastic structural modes are presented in Table 11.2. These modal frequencies are computed for the aircraft with no aerodynamics considered.

Table 11.2: Natural Frequencies of the Structural Finite Element Model of the Full Fuel F/A-18 SRA with Left Wing Modifications and No Wingtip Excitation System

Mode	Symmetric (Hz)	AntiSymmetric (Hz)
Wing 1st Bending	5.59	8.84
Fuselage 1st Bending	9.30	8.15
Stabilator 1st Bending	13.21	12.98
Wing 1st Torsion	13.98	14.85
Vertical Tail 1st Bending	16.83	15.61
Wing 2nd Bending	16.95	16.79
Wing Outboard Torsion	17.22	-
Fuselage 2nd Bending	19.81	18.62
Trailing Edge Flap Rotation	23.70	23.47
Fuselage Torsion	-	24.19
Launcher Rail Lateral	-	24.35
Stabilator Fore and Aft	28.31	28.58
Wing 2nd Torsion	29.88	29.93
Aileron Rotation	33.44	-
Aft Fuselage Torsion	-	37.80
Aileron Torsion	38.60	-
Wing Pitch	-	39.18
Wing 3rd Bending	43.17	-

The predicted flutter results for this aircraft are computed from the finite element model using the *p-k* method [57]. The doublet lattice and constant panel methods are used to compute the frequency-varying unsteady aerodynamic forces for several subsonic,transonic and supersonic Mach numbers.

Force matrices for Mach numbers, $M = .8, .9, .95, 1.1, 1.2, 1.4, 1.6$, are available. The unsteady aerodynamic forces are computed as a function of reduced frequency, k.

$$k = \omega \frac{\overline{c}}{2V} \tag{11.1}$$

The reduced frequency is a function of the true frequency, ω, the true velocity, V, and $\overline{c}$ the mean aerodynamic chord. Aerodynamic forces generated for 10 reduced frequency points between $k = .0001$ and $k = 4$ are sufficient for flutter margin computation for this aircraft. Figure 11.8 presents the magnitude and phase of a sample calculation of the unsteady aerodynamic forces as a function of reduced frequency. An additional 10 points computed between $k = .01$ and $k = 1$ demonstrate the function is smooth across frequency so the original set of 10 points is sufficient to describe the behavior of unsteady aerodynamic forces.

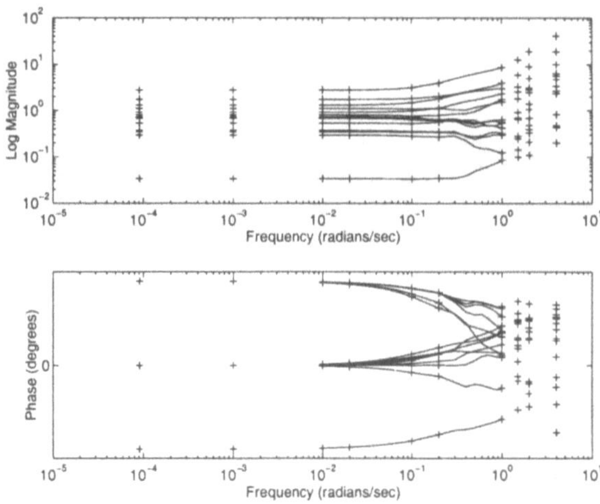

Fig. 11.8: Magnitude and Phase of the Frequency Varying Unsteady Aerodynamic Forces : 10 Points Between Reduced Frequency $k = .0001$ and $k = 4$ (+), 10 Points Between Reduced Frequency $k = .01$ and $k = 1$ (—)

The unsteady aerodynamic forces are fit to a finite-dimensional state-space system. The system identification algorithm is a frequency domain curve fitting algorithm based on a least squares minimization [93]. A separate system is identified for each column of the unsteady forces transfer function matrix. 4th order state-space systems are used for each column of the symmetric forces and 2nd order state-space systems are used for each column of the antisymmetric forces. These systems are combined to form a single multiple-input and multiple-output state-space model of the unsteady aerodynamics forces, previously designated $Q(s)$, with 56 states for the symmetric modes and 28 states for the antisymmetric modes.

The analytical aeroelastic model has inputs for symmetric and antisymmetric excitation forces. It is assumed the excitation force will be purely symmetric or antisymmetric. There are 10 outputs corresponding to the sensor locations on the aircraft.

11.4 Uncertainty Description

Noise and uncertainty operators are introduced to the linear aeroelastic model to account for variations between the analytical model and the actual aircraft. Standard analysis of the linear model is used to determine the framework for how uncertainty operators enter the system. Two uncertainty operators and a single noise input are used to describe the modeling uncertainty in the linear aeroelastic model. The magnitude of each uncertainty operator and the noise level is determined both from reasoning of the modeling process and analysis of the flight data [79].

An uncertainty operator, Δ_A, is associated with the state matrix of the F/A-18 linear model. This uncertainty models variations in both the natural frequency and damping values for each mode [6]. State matrix uncertainty can account for errors in coefficients of the equations of motion and changes in the aircraft dynamics due to parameter variations such as mass consumption during flight.

Δ_A is a structured diagonal matrix with real scalar parameters as elements. Separate elements are used to affect each modal response and time lag in the state matrix. The modal response uncertainty parameters are each repeated two times while each time lag uncertainty appears once on the diagonal.

Each repeated modal uncertainty parameter affects natural frequency and damping by allowing variation in the state matrix elements. Consider formulating the state matrix as block diagonal with a 2×2 block representing each mode. The real component of a modal eigenvalue, e_r, is the diagonal component of each block while the imaginary part, e_i, is arranged on the off-diagonal positions. Define A_i as the block determining the natural frequency, ω_i, and damping, ζ_i, of the ith mode.

$$A_i = \begin{bmatrix} e_r & e_i \\ -e_i & e_r \end{bmatrix} \tag{11.2}$$

The modal parameters that are observed with flight are the natural frequency computed as $\omega_i = \sqrt{e_r^2 + e_i^2}$ and damping computed as $\zeta_i = -e_r/\omega_i$. Of course, there are different blocks in the state matrix corresponding to different modes so the subscript indicates the parameters for the ith mode.

Scalar weightings, w_r and w_i, are used to affect the amount of uncertainty in each matrix element. The amount of variation in the matrix elements, and correspondingly the amount of variation in the natural frequency and damping, are determined by the magnitude of these scalar weightings. Define $\overline{e_r}$ and $\overline{e_i}$ as elements of the state matrix affected by an uncertainty δ.

$$\overline{e_r} = e_r(1 \pm w_r\delta) \tag{11.3}$$

$$\overline{e_i} = e_i(1 \pm w_i\delta) \tag{11.4}$$

Aeroelastic modes typically show low damping values caused by the real component being quite small as compared to the imaginary component. Since linear modeling techniques often identify the natural frequency better than the damping value, the weighting for the real component is expected to be larger than that for the imaginary component. This is reflected by the observed modal parameters in the flight data. The natural frequencies show variations of $\pm 5\%$ from the theoretical model while the uncertainty in the damping needs approximately $\pm 15\%$ to validate all the flight data. The scalar weightings are chosen accordingly.

$$w_r = .15 \tag{11.5}$$

$$w_i = .05 \tag{11.6}$$

The flight data is only able to determine uncertainty levels for the modal parameters of the experimentally observed modes. It is assumed the uncertainty levels in the unobserved modes should be consistent with these values. Parametric uncertainty is introduced for each modal block in the state matrix, affecting observed and unobserved modes, with the weighting values given above.

The single scalar blocks of Δ_A are normalized to one by weighting their effect on the time lags of the state matrix. Variations observed in the flight data is used to determine a weighting of $w_{lag} = .15$ is required to admit 15% variation in the time lags.

The second uncertainty operator, Δ_{in}, is a multiplicative uncertainty on the force input to the linear model. This uncertainty is used to cover nonlinearities and unmodeled dynamics. The linear model contains no dynamics above 40 *Hz* so the high frequency component of this operator will reflect this uncertainty. This operator is also used to model the excitation uncertainty due to the DEI exciter system. Analysis of the flight data indicates the input excitation signals rarely had the desired magnitude and phase characteristics that they were designed to achieve. The low frequency component of the input uncertainty reflects the uncertainty associated with the excitation system used to generate the flight data. The frequency-varying transfer function for weighting the input uncertainty is given as W_{in}.

$$W_{in} = 5\frac{s+100}{s+5000} \tag{11.7}$$

A noise signal is included with the sensor measurements. Knowledge of the aircraft sensors is used to determine that a level of 10% noise is possible in the measured flight data. An additional noise may be included on the force input due to the excitation system but it is decided the input multiplicative uncertainty is sufficient to describe this noise.

The block diagram for the aeroelastic model with the uncertainty operators is given in Figure 11.9. Note the weightings associated with Δ_A are actually internally scaling the F/A-18 operator and so are not explicit modeled.

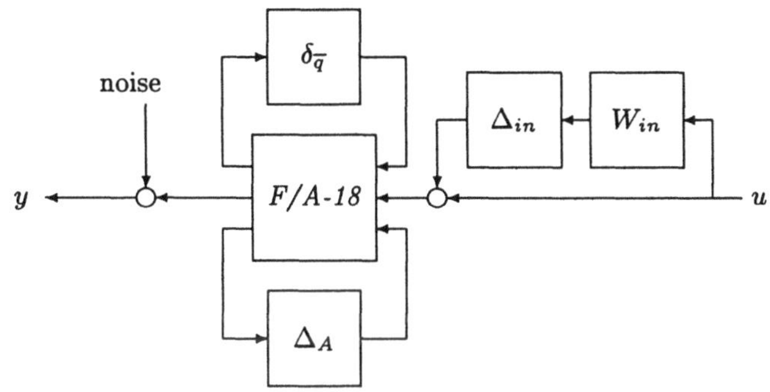

Fig. 11.9: F/A-18 Robust Stability Block Diagram

The flight data used to validate this uncertainty structure covers a large range of flight points. The entire set of 260 flight maneuvers throughout the flight envelope is considered. Using a single uncertainty description over the entire flight envelope may be conservative. Certainly it is reasonable to assume the linear models are more accurate at subsonic and supersonic than at transonic because of flow irregularities due to shock formation. Additionally, the flight data from the DEI exciter system should be better at subsonic speeds than at supersonic because of the degraded performance at higher dynamic pressures. However, it simplifies the analysis process to consider a single set of uncertainty operators. This process is equivalent to formulating the worst-case uncertainty levels at the worst-case flight condition and assuming that amount of uncertainty is possible for the remaining flight conditions.

11.5 Nominal and Robust Flutter Pressures

Flutter pressures are computed for a linear model with the associated modeling uncertainty structure using the μ-analysis method [74]. Linear systems for symmetric and antisymmetric structural modes are separated for ease of analysis. These systems can easily be combined and analyzed as a single system; however, eigenvector analysis would be required to distinguish which critical flutter modes are symmetric and which are antisymmetric. Each system contains the same number of structural modes, 14, and the uncertainty descriptions are identical for each linear model.

The system given in Figure 11.9 contains three uncertainty blocks. The parametric uncertainty covering variations due to dynamic pressure, $\delta_{\overline{q}}$, is a scalar parameter repeated 14 times, once for each elastic mode. The parametric uncertainty block affecting the modal parameters, δ_{modes}, is a diagonal matrix with dimension equal to the number of states. Separate scalars along the diagonal represent uncertainty in each elastic mode, each mode in the aerodynamic force approximation, and each lag term. The uncertainty parameters for the modes are repeated two times while the parameters for the lag terms are single scalars. Define δ_i as the ith uncertainty parameter for the system with n_m modes and n_l lag terms. The input multiplicative uncertainty block, Δ_{in}, is a scalar for this single input plant model since we are analyzing symmetric excitation separately from antisymmetric excitation.

The parametric uncertainty parameters represent changes in elements of the state-space model. The variation of $\delta_{\overline{q}}$ between ± 1 admits dynamic pressures between $0 \leq \overline{q} \leq 2\overline{q}_o$. Allowing the modal uncertainty parameters, $\delta_1, \ldots, \delta_{n_m}$ to vary between $\pm$ 1 allows 5% variation in the imaginary part of the natural frequency and 15% in the real part. This corresponds to approximately 5% variation in the natural frequency and 15% in the damping value of each mode. These parameters are real quantities. The multiplicative input uncertainty contains magnitude and phase information and is treated as a complex linear time-invariant uncertainty.

Nominal flutter pressures are initially computed by ignoring the modal and input uncertainties. μ is computed only with respect to the parametric uncertainty allowing a range of dynamic pressures to be considered. Robust flutter pressures are computed with respect to the structured uncertainty set, Δ, described above using the structured singular value. Traditional flutter pressures computed using the *p-k* method are presented with the nominal and robust flutter pressures computed with μ in Table 11.3

The nominal flutter dynamic pressures computed using the μ method can be directly compared with those computed using the traditional *p-k* method. Each of these flutter solutions are based on an analytical model with no consideration of modeling uncertainty.

The nominal flutter points for the symmetric modes match closely with the *p-k* method throughout the flight envelope. The subsonic and supersonic cases show an especially good correlation with the *p-k* flutter points. For each

Table 11.3: Nominal and Robust Flutter Pressures (lb/ft^2) Computed with *p-k* and μ Methods

	symmetric			antisymmetric		
Mach	$\bar{q}_{p-k}$	$\bar{q}_{nom}$	$\bar{q}_{rob}$	$\bar{q}_{p-k}$	$\bar{q}_{nom}$	$\bar{q}_{rob}$
.8	3360	3168	2909	4600	4593	3648
.9	2700	2706	2575	3150	3057	2944
.95	2430	2388	2329	2600	2751	2572
1.1	5400	5676	4120	5500	3265	2827
1.2	2469	2454	2327	2850	2893	2653
1.4	3528	3432	3034	4600	4439	4191
1.6	4470	4487	3996	5700	5870	4536

of these flight regions, the μ-analysis flutter dynamic pressures are nearly identical, within 1%, to the *p-k* method flutter dynamic pressures. The transonic case at $M = 1.1$, however, shows a slight difference between the two methods. The μ method computes a flutter point that is greater than the *p-k* method. In each Mach regime; subsonic,supersonic or transonic, the nominal flutter points are within 5% for the two methods.

The antisymmetric modes show a similar relationship between the flutter pressures computed with the μ and *p-k* methods. The subsonic and supersonic flutter points are within 5% for the two methods, but there is a greater deviation at the transonic condition. μ computes a flutter pressure at $M = 1.1$ that is 40% lower than the *p-k* method indicates.

The nominal flutter points for the μ and *p-k* methods show the greatest difference for both the symmetric and antisymmetric modes at the transonic case. The aerodynamics at $M = 1.1$ are more difficult to model accurately than at either subsonic or supersonic. Numerical sensitivity to representations of the unsteady aerodynamic forces causes differences in the nominal flutter pressure.

The robust flutter pressures computed using the μ method have lower dynamic pressures than the nominal pressures, which indicates the expected conservative nature of the robust computation. These new flutter points are *worst-case* values for the entire range of allowed uncertainty. The subsonic and supersonic flutter boundaries are not greatly affected by the uncertainty set. In each of these cases, the robust flutter point is within 10% of the nominal flutter point.

The flutter boundary at the transonic case, $M = 1.1$, demonstrates significant sensitivity to the modeling uncertainty. The robust flutter dynamic pressures are approximately 70% of the nominal flutter pressures. This is explained by considering the rapid transition of critical flutter boundaries near this region. The critical flutter frequencies and the flutter dynamic pressure widely vary between Mach numbers slightly lower and higher than transonic. The small amount of modeling uncertainty is enough to cause the worst-case flutter mechanism to shift and the plant assumes characteristics more consistent with a non-transonic regime.

The modal natural frequencies for the critical flutter modes are presented in Table 11.4. The frequencies computed using the *p-k* method and the μ-analysis method are close throughout the flight envelope for both the symmetric and antisymmetric modes. Frequencies for the robust flutter solutions are slightly different than the nominal flutter frequencies due to the modeling uncertainty which allowed 5% variation in the modal natural frequencies.

Table 11.4: Frequencies (em Hz) of Unstable Modes Associated with Nominal and Robust Flutter Pressures Computed with *p-k* and μ Methods

	symmetric			antisymmetric		
Mach	ω_{p-k}	ω_{nom}	ω_{rob}	ω_{p-k}	ω_{nom}	ω_{rob}
.8	8.2	7.6	7.7	9.0	9.1	9.1
.9	7.4	7.3	7.3	9.2	9.1	9.2
.95	6.8	6.9	6.9	9.1	9.2	9.2
1.1	12.1	13.2	13.0	28.6	28.0	28.3
1.2	26.5	27.4	27.4	26.9	28.9	28.9
1.4	28.1	28.1	28.1	30.4	31.7	31.8
1.6	28.9	30.1	30.1	32.8	32.3	32.1

Subcritical flutter pressures computed with the μ and *p-k* methods are presented in Table 11.5. Only nominal subcritical pressures are detected with μ since the robust pressures are always worst-case critical pressures.

Table 11.5: Subcritical Nominal Flutter Pressures (lb/ft^2) Computed with *p-k* and μ Methods

	symmetric		antisymmetric	
Mach	$\bar{q}_{p-k}$	$\bar{q}_{nominal}$	$\bar{q}_{p-k}$	$\bar{q}_{nominal}$
.9			4700 5300	4583 5093
.95	7450	6919		
1.1			6050	6001
1.2	5400	5003	8400	7943
1.4	8970	8959		
1.6	8400	8843		

μ-analysis computes subcritical flutter pressures within 10% of the *p-k* method for both the symmetric and antisymmetric modes. The μ method is even able to detect the subcritical flutter hump mode occurring for antisymmetric excitation at 0.9 Mach number.

11.6 Nominal and Robust Flutter Margins

The nominal and robust flutter pressures in Table 11.3 are used to compute flutter margins for a flight envelope. These flutter margins are determined by the distance between the flutter pressure and the dynamic pressure at the edge of the flight envelope at a particular Mach number.

Table 11.6 lists the flight conditions for several points above Mach 0.8 with the highest dynamic pressures in the defined flight envelope along with the nominal and robust flutter margins. These flutter margins all positive values indicating the envelope is free of flutter instabilities. The critical flutter pressures are simply the sum of the flight envelope dynamic pressures and the flutter margins.

Table 11.6: Nominal and Robust Flutter Margins (lb/ft^2) for a Flight Envelope Expressed as Allowable Increase in Dynamic Pressure Computed with *p-k* and μ Methods for Symmetric and Antisymmetric Modes

flight envelope		symmetric			antisymmetric		
Mach	$\bar{q}$	Γ_{p-k}	Γ_{nom}	Γ_{rob}	Γ_{p-k}	Γ_{nom}	Γ_{rob}
.8	948	2412	2220	1961	3652	3645	2700
.9	1200	1500	1506	1375	1950	1857	1744
.95	1337	1093	1051	992	1263	1414	1235
1.1	1491	3909	4185	2629	4009	1774	1336
1.2	1438	1031	1016	889	1412	1455	1215
1.4	1252	2276	2180	1782	3348	3187	2939
1.6	1027	3443	3460	2969	4673	4843	3509

The flutter margins are often expressed in terms of the percent change, Π, in knots of equivalent airspeed, KEAS, between the flight envelope and the flight condition corresponding to the critical flutter pressure. Table 11.7 lists these percentage values.

Table 11.7: Nominal and Robust Flutter Margins for a Flight Envelope Expressed as Percent of Allowable Increase in Knots of Equivalent Airspeed Computed with *p-k* and μ Methods for Symmetric and Antisymmetric Modes

flight envelope		symmetric			antisymmetric		
Mach	KEAS	Π_{p-k}	Π_{nom}	Π_{rob}	Π_{p-k}	Π_{nom}	Π_{rob}
.8	529	88	82	75	120	120	96
.9	595	50	50	46	62	59	56
.95	628	35	33	31	39	43	38
1.1	663	90	95	66	92	48	37
1.2	652	31	31	26	41	41	35
1.4	608	67	65	55	91	88	82
1.6	550	108	109	97	135	139	109

A flutter margin yielding a percentage $\Pi = 15$ for allowable increase in knots of equivalent airspeed is commonly used to determine an operable flight envelope. The flutter margins given in Table 11.7 demonstrate the indicated flight envelope for the F/A-18 SRA is acceptable since nominal and robust flutter margins are all greater than 15% for symmetric and antisymmetric modes. The smallest flutter margin occurs for the symmetric mode dynamics at Mach 1.2 which presents a robust flutter margin of only 26%; however, even this condition is greater than the desired 15% margin.

The dynamic pressures at which flutter occurs are converted into altitudes, commonly known as matched-point solutions, using standard atmospheric equations. These altitudes are plotted for the symmetric modes in Figure 11.10 and for the antisymmetric modes in Figure 11.11. The flight envelope of the F/A-18 is shown on these plots along with the required 15% flutter boundary for military aircraft.

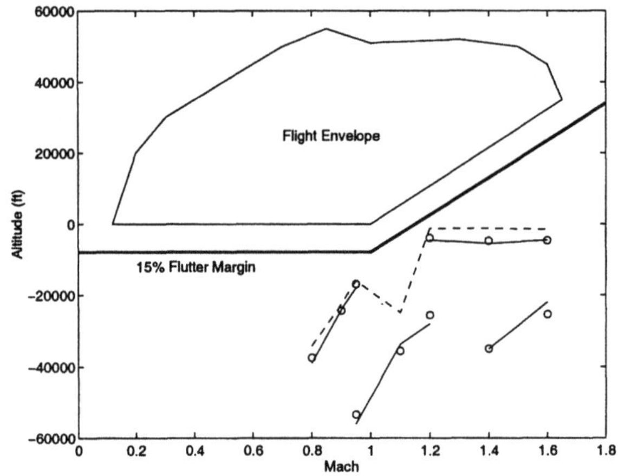

Fig. 11.10: Nominal and Robust Flutter Points for Symmetric Modes : nominal *p-k* margin (—), nominal μ margin (o), robust μ margin (- - -)

Figures 11.10 and 11.11 use several short solid lines to indicate the *p-k* flutter solutions throughout the flight regime. Each of these short solid lines represents the flutter points due to a specific mode. Flutter points for the symmetric modes given in Figure 11.10 show four solid lines indicating three different critical flutter modes for the considered range of Mach numbers along with a subcritical flutter mode occurring at supersonic Mach numbers. The antisymmetric modes show the onset of flutter from three different critical modes along with three subcritical flutter modes throughout the flight envelope in Figure 11.11. The frequencies of the critical flutter modes can be found in Table 11.4.

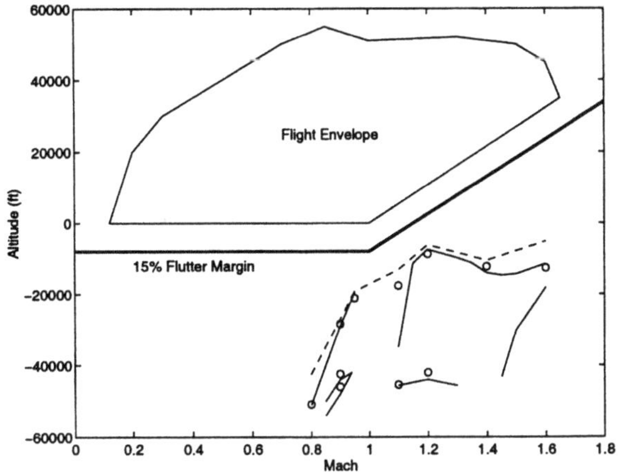

Fig. 11.11: Nominal and Robust Flutter Points for AntiSymmetric Modes : nominal p-k margin (—), nominal μ margin (o), robust μ margin (- - -)

The subsonic flutter altitudes for symmetric and antisymmetric modes demonstrate a similar characteristic. The nominal flutter boundary shows a significant variation from Mach number $M = .8$ to $M = .95$ caused by sensitivity to Mach number for the dynamics associated with the critical flutter mode. The robust flutter boundary indicates the sensitivity of the plant to errors and the worst-case perturbation. The higher altitude for the nominal flutter boundary at Mach number $M = .81$ than for Mach number $M = .80$ is reflected in the large conservatism associated with the robust flutter boundary. Similarly, slight variation of Mach number near $M = .95$ is not expected to decrease the nominal flutter margin so there is less conservatism associated with the robust flutter boundary.

An interesting trend is noticeable for the symmetric mode robust flutter points in Figure 11.10 at the supersonic Mach numbers. The flutter mechanism results from the same modes from $M = 1.2$ to $M = 1.6$ with some increase in frequency. Similarly the altitudes of the nominal flutter margins show little change for these Mach numbers. The aeroelastic dynamics associated with the critical flutter mode are relatively unaffected by the variation of Mach over this range and consequently each flutter boundary has the same sensitivity to modeling errors.

The robust flutter margins for the antisymmetric modes at supersonic Mach numbers show a slightly different behavior than the symmetric mode flutter margins. The flutter instability is again caused by a single mechanism from $M = 1.2$ to $M = 1.6$ with similar frequency variation as symmetric. The robust flutter margins demonstrate a similar sensitivity to modeling errors at $M = 1.2$ and $M = 1.4$ but at $M = 1.6$ a greater sensitivity is shown.

The greater conservatism at $M = 1.6$ may indicate impending transition in flutter mechanism to the subcritical mode at slightly higher Mach number.

The dark solid line on Figures 11.10 and 11.11 represents the required boundary for flutter points. All nominal and robust flutter points lie outside this region indicating the flight envelope should be safe from flutter instabilities. The robust flutter boundaries computed with μ indicate there is more danger of encountering flutter than was previously estimated with the *p-k* method. In particular, the robust flutter margin for symmetric excitation at Mach $M = 1.2$ lies considerably closer to the boundary than the *p-k* method indicates.

11.7 Computational Evaluation

The μ analysis method of computing flutter margins presents significant analytical advantages due to the robustness of the resulting flutter margin, but it also has several computational advantages over the *p-k* method. The μ algorithm requires a single linear aeroelastic plant model at a given Mach number to compute critical and subcritical flutter margins. An entire set of flutter margins may be easily generated using a standard engineering workstation in a few minutes using commercially available software packages [5].

The *p-k* method is an iterative procedure that requires finding a matched-point solution [57]. The aircraft is analyzed at a particular Mach number and air density. The airspeed for these conditions resulting in a flutter instability is computed. This airspeed, however, often does not correspond to the unique airspeed determined by that Mach number and air density for a standard atmosphere. Various air densities are used to compute flutter solutions and the corresponding air speeds are plotted. An example of an air speed plot for flutter is given in Figure 11.12.

The vertical lines in Figure 11.12 represent two antisymmetric modes that may flutter at Mach M=1.4. The *p-k* method computes a flutter solution at the airspeed indicated where the modal line crosses the standard atmosphere curve. This flutter solution is difficult to compute from only a few air density computations. Typically several air densities are used to compute air speed flutter solutions and a line is extrapolated between the points to determine the matched-point solution at the standard atmosphere crossing point. The nonlinear behavior shown for mode 1 near the standard atmosphere crossing point indicates an accurate flutter boundary would be extremely hard to predict unless many solutions are computed near the true matched-point solution.

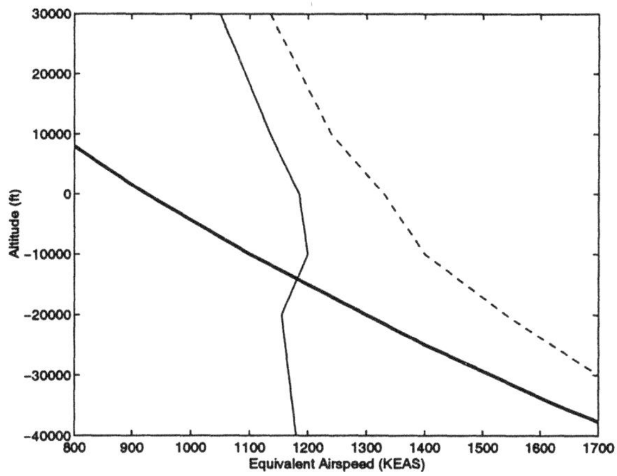

Fig. 11.12: AntiSymmetric *P-K* Flutter Solutions for Mach 1.4 : Wing 2nd Torsion mode (—), Trailing Edge Flap Rotation mode (- - -), also shown for reference is the standard atmosphere curve (___)

The *p-k* method also may have difficulty predicting the subcritical flutter margins. The second mode in Figure 11.12 may or may not intersect the standard atmosphere curve. More computational analysis is required to determine the behavior of this mode at higher airspeeds. The μ-analysis method accurately detects both the critical and subcritical flutter margins without requiring expensive iterations.

CHAPTER 12

ROBUST AEROSERVOELASTIC STABILITY OF THE F / A-18 HARV

12.1 Sensing and Control Elements

The F/A-18 High Alpha Research Vehicle (HARV) was flown at NASA Dryden Flight Research Center to conduct flight tests at high angle-of-attack flight conditions. This paper considers the F/A-18 HARV with a thrust vectoring system to direct the flow out of the engine nozzles. This aircraft, shown in Figure 12.1, is sometimes denoted as the HARV-TVCS to emphasize the thrust vectoring control system, but for convenience the additional TVCS notation will not be used here.

Fig. 12.1: F/A-18 HARV with Thrust Vectoring in Flight

The F/A-18 HARV underwent several structural modifications to enable controlled high angle-of-attack flight with the thrust vectoring system. These modifications include additions such as three vanes in each engine exhaust and associated ballast in the forward fuselage to maintain a desired center of gravity position; a research flight control system; an inertial navigation system for angle-of-attack and sideslip rate computation at high angle-of-attack flight conditions; wingtip vanes for sensors and air data probes; and additional instrumentation. The F/A-18 HARV has significantly different structural dynamic properties from a standard F/A-18 aircraft as a result of these modifications.

The control system consists of a basic flight control system and a research flight control system (RFCS). The additional RFCS controller is used to command the thrust vectoring at high angle-of-attack flight conditions [49]. Separate state-space controllers are developed for longitudinal and lateral-directional control with distinct gain scheduling laws determining the controller gains at different conditions within the flight envelope [99]. These controllers are discrete-time systems which operate at an 80 *Hz* update rate.

The longitudinal controller, K_{long}, is a 15 state system which operates on the feedback measurements and returns the controller commands given in Table 12.1.

Table 12.1: Feedback Measurements and Controller Commands for the Thrust Vectoring Longitudinal Controller K_{long}

Feedback Measurements			Controller Commands		
pitch rate	q	deg/sec	symmetric stabilator	δ_e	deg
alpha derivative	$\dot{\alpha}$	deg/sec	leading edge flap	δ_{lef}	deg
alpha	α	deg	trailing edge flap	δ_{tef}	deg
			pitch thrust vectoring	δ_{tv}^{p}	deg

The lateral-directional controller, K_{lat}, is a 29 state system which operates on the feedback measurements and returns the controller commands given in Table 12.2.

Table 12.2: Feedback Measurements and Controller Commands for the Thrust Vectoring Lateral-Directional Controller K_{lat}

Feedback Measurements			Controller Commands		
roll rate	p	deg/sec	differential stabilator	δ_{de}	deg
slide slip rate	$\dot{\beta}$	deg/sec	aileron	δ_a	deg
yaw rate	r	deg/sec	differential leading edge flap	δ_{dlef}	deg
lateral acceleration	n_y	g	differential trailing edge flap	δ_{dtef}	deg
			rudder	δ_{rud}	deg
			yaw thrust vectoring	δ_{tv}^{y}	deg

The flight envelope for the F/A-18 HARV is a small subsonic box within the flight envelope of a standard F/A-18 as given in Figure 12.2.

12.2 Analytical Model

A finite element model of the F/A-18 HARV is used to generate a linear time-invariant model to represent the structural dynamics. This model contains 10

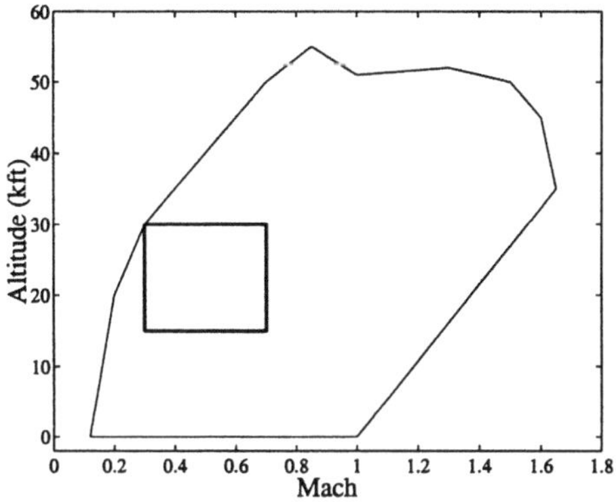

Fig. 12.2: Flight Envelopes of a F/A-18 Aircraft : Standard Production F/A-18 (—) and F/A-18 HARV with RFCS

elastic symmetric and 10 elastic antisymmetric structural modes along with 6 rigid-body modes. High order models are used to represent the actuator dynamics and add 90 states to the model. There are additionally 12 control surface modes. Including the state-space representation of the unsteady aerodynamic forces introduces 20 states for the symmetric modes and 20 states for the antisymmetric modes.

Table 12.3 lists some elastic modal natural frequencies in this state-space model for the F/A-18 HARV at heavyweight condition with full fuel. There are also several modes associated with the thrust vectoring vane dynamics that have natural frequencies slightly above 20 *Hz* but they are not confidently accepted because of modeling difficulties and so are not included in Table 12.3.

Separate models represent the closed-loop aeroservoelastic dynamics at the corners of the research flight envelope shown in Figure 12.2. These corners represent the most extreme flight conditions with respect to Mach and dynamic pressure so the stability margins are anticipated to be smallest at these conditions. Additionally, only decreasing values of dynamic pressure will be considered since any aeroservoelastic instabilities are anticipated to be associated with low dynamic pressures [22].

Stability margins are computed by considering perturbations to dynamic pressure for the models with constant controller gains. This is valid because RFCS gain are scheduled within the envelope but are fixed to their values from the closest edge of the envelope when flying beyond the design ranges to lower dynamic pressures. Thus, the stability margins indicate the flight conditions nearest the envelope for which instabilities may be encountered.

Table 12.3: Natural Frequencies in *Hz* of the Structural Finite Element Model of the Heavyweight Full Fuel F/A-18 HARV with Thrust Vectoring System

Mode	Symmetric	AntiSymmetric
Wing 1st Bending	5.75	8.61
Fuselage 1st Bending	7.45	6.86
Wing 1st Torsion	11.64	12.02
Stabilator 1st Bending	13.68	13.57
Fin 1st Bending	15.92	15.56
Fuselage 2nd Bending	15.34	21.01
Wing 2nd Bending	17.05	-
Wing Fore-Aft	18.19	14.99
Fuselage 1st Torsion	-	18.76
Vane Rotation	22.10	22.10
Inboard Flap Rotation	23.50	23.02
Wingtip Torsion	27.47	-

12.3 Uncertainty Description

An uncertainty description, $\mathbf{\Delta}$, is formulated that uses 3 operators to describe errors in a HARV analytical model. A complex operator, Δ_{in}, is a multiplicative uncertainty in the control inputs to the plant and accounts for actuator errors and unmodeled dynamics. Another complex operator, Δ_{add}, relates the control inputs to the feedback measurements to account for uncertainty in the magnitude and phase of the computed plant responses. The remaining uncertainty operator, Δ_A, is a real parametric uncertainty affecting the modal parameters of the open-loop state matrix to describe errors in natural frequency and damping parameters.

The block diagram for robust stability analysis of the F/A-18 HARV aeroservoelastic dynamics is shown in Figure 12.3. This Figure includes an operator, $\delta_{\overline{q}}$, that affects the nominal dynamics to describe changes in flight condition and is used to interpret μ as a stability margin. Additional operators, W_{add} and W_{in}, are shown as weightings to normalize the frequency varying uncertainty operators, Δ_{add} and Δ_{in}. The system model also contains an extraneous 5% noise signal corrupting the sensor measurements.

This uncertainty description is used to describe errors in the models from each corner of the flight envelope. Two sets of magnitudes are generated with one set describing the amount of uncertainty in the symmetric dynamics model and the other set describing the amount of uncertainty in the antisymmetric dynamics model. In this sense, $\mathbf{\Delta}$ is a worst-case uncertainty description that contains the largest errors for the symmetric or antisymmetric models at any point in the research flight envelope.

The magnitudes of the uncertainty are chosen from analysis of measured flight data using a model validation procedure based on the condition in Theorem 7.1.2. This procedure determines if the measured flight data could have been generated by some plant comprised of the nominal model and

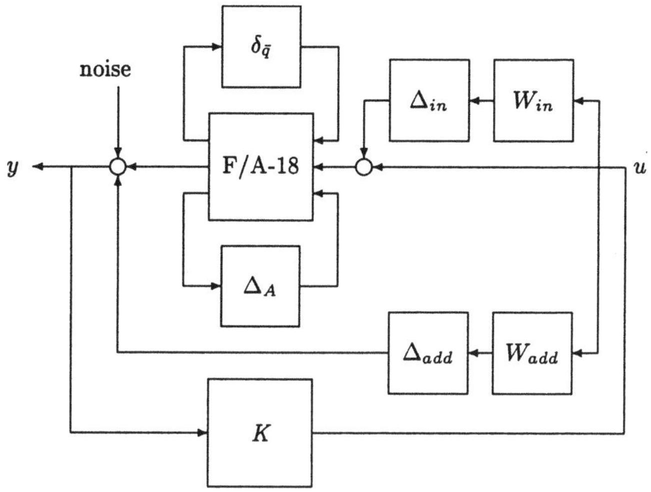

Fig. 12.3: Closed-Loop F/A-18 HARV Model with Uncertainty

a perturbation described by the uncertainty [75]. Values of uncertainty are iteratively tested until the family of plants described by Figure 12.3 has a range of responses which can sufficiently bound the measured flight data transfer functions.

The modal parametric uncertainty for the antisymmetric dynamics model is determined by considering flight data sets generated in response to lateral-directional excitation. Consider the transfer functions from yaw thrust vectoring command (δ_{tv}^{y}) to normal acceleration (n_y) for several analytical models and data taken at Mach 0.3 and 30000 *feet* shown in Figure 12.4.

The dashed line in Figure 12.4 represents the nominal model transfer function and clearly does not match the solid line showing the experimental flight data transfer function. The two thick solid lines in this Figure demonstrate the range of transfer functions that can be generated by introducing modal parametric uncertainty. Each thick line corresponds to a different value of the modal perturbation and it can be seen that both perturbation values are needed to validate the model. Introducing one perturbation ensures the model closely matches the data near 16 *Hz* while the other perturbation is required near 18 *Hz*.

Separate modal parametric uncertainty levels are chosen for each mode of the open-loop state matrix to reflect different levels of accuracy. Multiple flight data sets are analyzed and the largest errors observed between theoretical and experimental modal parameters are used to determine the uncertainty levels. The nominal modal parameters and the amount of variation admit-

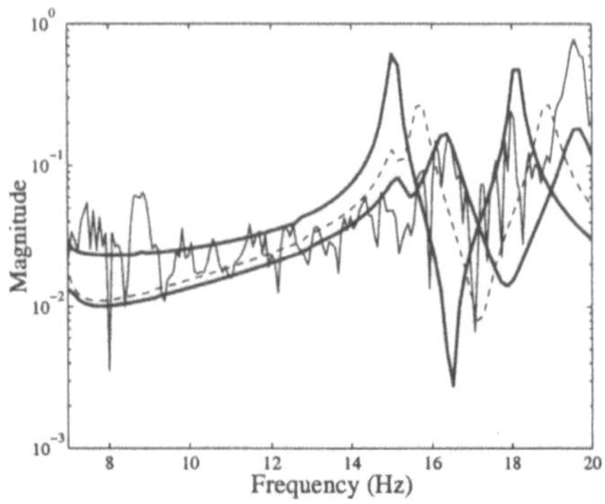

Fig. 12.4: Transfer Functions from δ^y_{tv} to n_y : experimental flight data (—), nominal model (- - -), perturbed models (—)

ted by the parametric uncertainty for the antisymmetric dynamics model are given in Table 12.4.

Table 12.4: Modal Parameters and Uncertainty Variations for Antisymmetric Model

mode	ω (Hz)	ζ
Fuselage 1st Bending	$6.85 \pm .07$	$.012 \pm .006$
Wing 1st Bending	$8.96 \pm .18$	$.006 \pm .004$
Wing 1st Torsion	$12.84 \pm .13$	$.011 \pm .006$
Wing Fore-Aft	$15.69 \pm .63$	$.010 \pm .007$
Fuselage 1st Torsion	$18.86 \pm .76$	$.010 \pm .005$

The amount of variation needed to describe modal parameter errors is fairly significant for all modes. The wing fore-aft and fuselage 1st torsion modes have properties which are particularly poorly modeled so there is up to 4% error in natural frequency and 70% error in damping. The remaining modes are relatively better modeled and have only 1% error in natural frequency but still require at least 50% error in damping.

A similar procedure is performed to compute the modal parametric uncertainty levels for the symmetric dynamics model. Transfer functions from the pitch thrust vectoring command (δ^p_{tv}) to normal acceleration (n_z) are used to observe errors between the model and the measured flight data. The nominal modal parameters and the amount of variation admitted by the generated parametric uncertainty are given in Table 12.5.

The amount of variation introduced to the modal parameters is again quite significant. The natural frequency variations are larger for the symmetric model than for the antisymmetric model with the largest error of 8% for

Table 12.5: Modal Parameters and Uncertainty Variations for Symmetric Model

mode	ω (Hz)	ζ
Fuselage 1st Bending	$7.55 \pm .15$	$.011 \pm .004$
Wing 1st Bending	$5.92 \pm .18$	$.013 \pm .005$
Wing 1st Torsion	$12.06 \pm .97$	$.013 \pm .002$
Wing 2nd Bending	$16.02 \pm .48$	$.011 \pm .004$
Wing Fore-Aft	$18.12 \pm .69$	$.011 \pm .005$

the wing fore-aft mode. The damping variations, however, are smaller for the symmetric model than for the antisymmetric model with errors in damping between 15% and 45% for all modes.

The weighting functions for the input multiplicative and additive uncertainties are chosen to account for any errors between the model and the flight data that can not be covered by the parametric modal uncertainty.

$$W_{in} = 10\frac{s+100}{s+5000}$$

$$W_{add} = .02$$

12.4 Stability Margins

Stability margins are computed for the F/A-18 HARV system in Figure 12.3 to determine the lowest dynamic pressures at which an aeroservoelastic instability occurs. The stability margin, Γ, computed from the μ method indicates the biggest reduction in dynamic pressure that may be considered before the onset of an ASE instability.

Nominal stability margins are computed for the aeroservoelastic model by computing the smallest destabilizing perturbation to dynamic pressure for the nominal model. μ is computed for this model to measure robustness with respect to the $\delta_{\bar{q}}$ operator in Figure 12.3 but ignoring the Δ uncertainty operators.

Robust stability margins are computed from μ analysis of the model in Figure 12.3. The smallest destabilizing perturbation to dynamic pressure, Γ, is computed such that some plant within the family of plants becomes unstable. Thus, the ASE dynamics are robust to the set of modeling errors, Δ, for dynamic pressures less than Γ away from the nominal flight condition.

Table 12.6 lists the nominal and robust stability margins for the longitudinal model describing the symmetric dynamics. The nominal margins are all significantly large to indicate a reduction of more than 2000 lb/ft^2 is required before the nominal dynamics become unstable. For each of these cases, the nearest instability is actually at a negative dynamic pressure, which is

Table 12.6: Nominal and Robust Stability Margins for Symmetric Dynamics Model with Longitudinal Controller K_{long}

Flight Condition		Nominal Stability Margin		Robust Stability Margin	
Mach	altitude	Γ	$\omega_{unstable}$	Γ	$\omega_{unstable}$
0.3	30 *kft*	-2523 lb/ft^2	4.8 *Hz*	-132 lb/ft^2	11.5 *Hz*
0.7	30 *kft*	-2710 lb/ft^2	4.8 *Hz*	-262 lb/ft^2	15.6 *Hz*
0.3	10 *kft*	-2447 lb/ft^2	4.8 *Hz*	-174 lb/ft^2	11.5 *Hz*
0.7	10 *kft*	-3153 lb/ft^2	4.8 *Hz*	-422 lb/ft^2	15.4 *Hz*

physically unrealistic, to indicate the nominal model with the longitudinal controller is effectively free from ASE instabilities.

The robust stability margins for the longitudinal model are lower than the nominal margins because of the conservatism introduced by the uncertainty; however, they are still significantly large. These margins demonstrate each plant model is robust to the indicated amount of modeling uncertainty despite reductions in dynamic pressure of at least 130 lb/ft^2 which generally corresponds to a negative dynamic pressure.

Table 12.6 also demonstrates the frequencies of the unstable modes associated with the robust stability margins are different than the frequency associated with the nominal margins. The critical mode for the nominal margins is the wing fore-aft mode, but for the robust margins the critical instability of the models at Mach 0.3 is the wing torsion mode and the critical instability of the models at Mach 0.7 is the fuselage bending mode. This shift in critical modes results from the inclusion of uncertainty that allows a variation to the wing torsion and fuselage dynamics that becomes unstable before the wing fore-aft mode.

Table 12.7 gives the nominal and robust stability margins for the lateral-directional model. Margins are computed only for plant models at the right side of the research flight envelope with Mach 0.3 because controller gains for K_{lat} were not available for the Mach 0.7 plant models.

Table 12.7: Nominal and Robust Stability Margins for Antisymmetric Dynamics Model with Lateral-Directional Controller K_{lat}

Flight Condition		Nominal Stability Margin		Robust Stability Margin	
Mach	altitude	Γ	$\omega_{unstable}$	Γ	$\omega_{unstable}$
0.3	30 *kft*	-268 lb/ft^2	14.8 *Hz*	-4 lb/ft^2	15.4 *Hz*
0.3	10 *kft*	-757 lb/ft^2	13.7 *Hz*	-17 lb/ft^2	8.9 *Hz*

These nominal margins are noticeably lower for this model as compared to the symmetric dynamics model; however the flight envelope is still clear of nominal ASE instabilities. The smallest stability margins for the antisymmetric dynamics is 268 lb/ft^2 which implies the nearest instability to the flight envelope is actually at a negative dynamic pressure and is physically unrealizable.

Robust stability margins are also computed for the antisymmetric dynamics models with respect to the uncertainty description in Figure 12.3. These margins, given in Table 12.7, are quite small and indicate the stability margins of the antisymmetric model are quite sensitive to modeling errors. Indeed, the lateral-directional controller may incur an aeroservoelastic instability if the aircraft reduces the dynamic pressure by as little as 4 lb/ft^2 from the top left corner of the research flight envelope.

The robust stability margins are considerably less than the nominal stability margins and indicate the importance of considering uncertainty. The nominal dynamics did not indicate any potential ASE problems but a robust analysis clearly shows the stability margins are strongly affected by errors in the model. This paper utilized a worst-case uncertainty description based on several flight data sets; however, a wavelet filtering technique has also been developed to reduce the conservatism in the uncertainty description and generate more confident stability margins [23, 24].

CHAPTER 13
ON-LINE ANALYSIS DURING A FLIGHT TEST

13.1 Flutterometer

Flight flutter testing is the process of envelope expansion that determines a range of flight conditions within which an aircraft is safe from aeroelastic instabilities. This testing, which must be done for new and reconfigured aircraft, incurs dramatic time and costs because of the danger associated with encountering unpredicted instabilities. A particular concern is the onset of explosive flutter that is characterized by a sudden loss of stability between test points despite only a small change in flight conditions [13].

Traditional methods of flight flutter testing analyze system parameters, such as damping levels, that vary with flight condition to monitor aircraft stability [66]. A real-time method to estimate the damping levels was developed based on a recursive prediction-error method [133]. This method was extended to improve the estimates by considering an extended Kalman filter in the formulation [115]. On-line methods using both time-domain and frequency-domain characteristics of turbulence response data have also been formulated to estimate dampings [116]. These methods monitored stability at test points but they were of limited usefulness for predicting the onset of flutter because damping may be highly nonlinear as flight conditions vary so damping trends may indicate stability despite proximity to an explosive flutter condition. An alternative eigenspace method was formulated based on orthogonality between eigenvectors, but this method uses a parameter that, similar to damping, indicates stability and may vary nonlinearly with flight condition [2].

The concept of a flutter margin was introduced in conjunction with a method that could predict the onset of flutter by analyzing data at subcritical airspeeds [146]. This method considered the interaction of two modes in the flutter mechanism to formulate a stability parameter that varied quadratically with dynamic pressure. This flutter margin technique was extended to consider several modes interacting as the flutter mechanism and demonstrate a prediction method for higher-order instabilities [106, 107, 62]. These flutter margins have been used for wind tunnel and flight test programs [10, 65]; however, the method is of limited applicability for general flight flutter testing because the assumptions of few modes coupling and the requirement to observe those modes may be too restrictive.

Stability parameters were also introduced as flutter margins that consider an autoregressive moving average process to describe the aeroelastic dynamics. One parameter was based on determinants from a stability criterion for discrete-time systems that are excited by random turbulence [87, 88]. A similar parameter was developed by extending the determinant method to consider short data segments with assumptions of local stationarity [89]. Another extension to this method derived a similar stability parameter but relaxed the requirements for stationarity [131]. These flutter margins can be applied to complex systems and require only turbulence for excitation; however, the flutter boundary is computed by extrapolating a nonlinear function and may be misleading.

The μ method is extended to consider on-line estimation of flutter margins during a flight test by incorporating flight data from a series of test points to compute flutter margins as the envelope is expanded [76]. The flight data is not used to estimate model parameters or identify a transfer function; rather, the data is only used to update the uncertainty description for the theoretical model. This approach avoids several difficulties in trying to estimate a high-order model from flight data that has low signal-to-noise ratio [95], but still accounts for time-varying dynamics by updating the uncertainty description to describe changing errors between the aircraft and the nominal model.

This method is advantageous to other flight flutter test methods because the stability parameter, μ, is essentially linear with changes in flight condition so instabilities can be accurately predicted. Also, this method can easily consider realistically high-order models and does not make assumptions about the number or type of modes that interact as the flutter mechanism.

The concept of a flutterometer is introduced as a flight test tool to indicate the proximity of a flutter condition. This tool is based on the on-line μ method and tracks the time-varying flutter margins as the envelope is expanded. A simulated flight test of an F/A-18 is used to demonstrate the advantages of the flutterometer as compared to traditional tracking of damping. The flutterometer continually provides accurate information about the flutter margin whereas the damping trends do not indicate an impending instability until the flutter margin is very small.

The flutterometer is defined as a tool that predicts flutter margins during a flight test. This tool can be immediately realized by on-line computing of robust flutter margins. The resulting flutterometer indicates the distance to flutter as determined by the largest increase in dynamic pressure for which the model is robustly stable. The displays used by this tool to relate information to pilots and engineers is not considered here; rather, a simulation is presented to demonstrate implementation and data analysis issues.

Traditional flight flutter testing for envelope clearance proceeds with several repeated operations [66]. The aircraft is trimmed at a stable flight condition on the edge of the envelope where measured aeroelastic responses are analyzed to estimate damping and modal characteristics. If the estimates do not indicate the imminent onset of flutter, then the envelope is slightly expanded and the procedure repeats until flutter or a performance limit is reached.

A flutterometer can easily be implemented on-line for such a flight test program by computing robust flutter margins at each test point. The procedure involves updating the model using current flight data, deriving a valid uncertainty description, and computing μ with respect to that uncertainty to determine the robust flutter margin. The flutterometer is updated at each test point to reflect the new robust flutter margin that is computed.

Many flight analysis facilities are increasing computational resources such that an on-line flutterometer can be implemented with a reasonable update rate that will not necessarily increase flight time. The accessibility of multiple computers will further assist the flight test analysis such that the flutterometer does not replace tracking of damping trends; rather, this tool augments traditional information used for envelope expansion. Also, flutterometers can be implemented using different data analysis approaches to compare flutter margins corresponding to local and global approaches on separate computers to provide additional information about aeroelastic stability.

The flutterometer described here makes direct use of the linear dependence of the aeroelastic dynamics on dynamic pressure. The μ method uses this dependence to compute robust flutter margin by considering linear perturbations to dynamic pressure. Thus, the flutterometer computes flutter margins in terms of dynamic pressure. A flutterometer has been formulated for flight tests that vary Mach but keep constant dynamic pressure; however, this paper will only consider flight tests that keep Mach constant and vary dynamic pressure.

13.2 Aircraft Model

A flight flutter test is simulated to demonstrate the on-line implementation of a flutterometer and compare with a traditional tracking of damping trends. The simulation uses models that are developed to represent the aeroelastic dynamics at Mach 1.2 of the F/A-18 Systems Research Aircraft (SRA) that is described in Chapter 11.

There are three linear state-space models used in the simulation.

P_{true}	: true aircraft dynamics
P_{nom}	: nominal theoretical model
P_{rob}	: robust theoretical model

P_{true} is assumed to represent the true F/A-18 SRA aeroelastic dynamics at Mach 1.2 and the flight test simulates its envelope expansion. This model is derived from a finite-element model that was used for flutter analysis of the F/A-18 SRA by the μ method and the *p-k* method [75, 79]. Of course, this model is only a representation of the physical aircraft but it can be used as a truth model for the purposes of the simulation. The flutterometer algorithm is generated without any knowledge of P_{true} and must use flight data that is simulated as responses to learn about the true dynamics.

The mass of P_{true} is time-varying throughout the simulation and represents the effect of fuel burning at the supersonic flight conditions. The mass is initially 95% of the full-fuel heavyweight mass at the start of the simulation and decreases to 90% of the heavyweight mass after 20 minutes of flight time. These values are chosen to be representative of reasonable fuel variations that are observed in flight.

A flutter analysis of P_{true} indicates that the trailing edge flap mode is the critical flutter mode. The dynamic pressure at which flutter occurs for this model at Mach 1.2 varies from $\overline{q} = 2360\ lb/ft^2$ for P_{true} at 95% of heavyweight mass to $\overline{q} = 2255\ lb/ft^2$ for P_{true} at 90% of heavyweight mass. The natural frequency of this mode remains relatively unchanged near 27.36 Hz despite the 5% variation in mass. The flight data analysis for the simulation will focus on this critical mode, although the flutterometer actually considers the dynamics over a broad frequency range.

The nominal model, P_{nom}, is used as the theoretical model of P_{true}. It is assumed the actual dynamics of P_{true} are not known; instead, the dynamics of P_{nom} are used to predict flutter margins. P_{nom} has 84 states that result from 14 structural modes and an additional 56 states used to model the unsteady aerodynamic forces. This model is formulated in the μ framework and includes the perturbation to dynamic pressure, $\delta_{\overline{q}}$, that allows a range of flight conditions to be considered.

There are several errors in the nominal model such that P_{nom} is not a completely accurate representation of P_{true}. These errors include incorrect structural damping and mass matrices. The damping of P_{nom} is too high and is 12% greater than the damping of P_{true}. Also, the mass of P_{nom} is constant at the value of the full-fuel heavyweight condition and does not account for fuel burn during flight. These errors are chosen because of specific variations between models and flight data noted from a previous F/A-18 flutter project; however, they represent typical errors that are often found in aeroelastic models [66].

The robust model, P_{rob}, is comprised of the nominal model, P_{nom}, and an associated uncertainty set to describe errors and unmodeled dynamics. This model uses operators, Δ_A and Δ_{in}, and an extraneous noise affecting the sensors to describe the uncertainty. Also, P_{rob} is formulated in the μ framework and includes the perturbation, $\delta_{\overline{q}}$, to affect dynamic pressure and describe changes in the aircraft dynamics associated with changes in flight condition.

The uncertainty operator Δ_A affects the modal parameters of the state matrix and accounts for errors in natural frequencies and dampings. This operator is a real diagonal matrix with separate scalar elements affecting the parameters of each mode. The interconnection between Δ_A and P_{nom} is initially weighted to allow 5% uncertainty in each natural frequency and 15% uncertainty in each damping.

The uncertainty operator Δ_{in} is a complex multiplicative uncertainty affecting the excitation force. A weighting function, W_{in}, reflect the frequency-varying levels of multiplicative uncertainty such that there is little error at low frequency but there is significant error above 40 *Hz* because of unmodeled dynamics.

$$W_{in} = 5\frac{s+100}{s+5000}$$

The block diagram for P_{rob} in the μ framework with uncertainty operators is given in Figure 13.1.

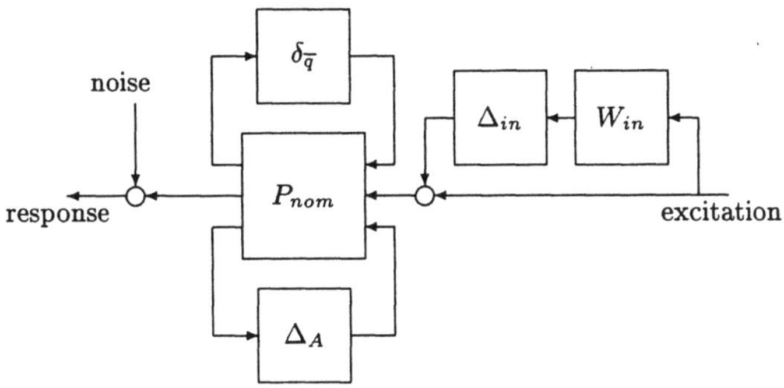

Fig. 13.1: Robust Model, P_{rob}, of the F/A-18 SRA

13.3 Simulated Flight Test

The simulation is designed to model the procedures used in a real flight flutter test. For ease of presentation, only the simulated flight test to determine the flutter boundary at Mach 1.2 for the symmetric modes is discussed. Flutter clearance of the entire flight envelope including antisymmetric modes is a straightforward extension to this simulation.

The flight test procedure expands the envelope by performing several operations at a series of test points [66].

1. decrease altitude to raise $\bar{q}$ by 100 *lb/ft*^{2}
2. measure symmetric response data
3. determine an uncertainty set such that the flight data does not invalidate the model
4. compute modal damping levels
5. compute robust flutter margin

The simulation is initialized at the dynamic pressure of $\overline{q} = 100\ lb/ft^2$ and altitude of 68.8 *kft* and the aircraft steadily decreases altitude until reaching the first test point at $\overline{q} = 700\ lb/ft^2$ and 27.7 *kft*. Further test points occur at intervals of 100 lb/ft^2 until the onset of a flutter condition. Naturally during a real flight test the aircraft would not purposely approach a flutter margin so closely, but encountering the flutter condition serves to demonstrate the accuracy of the computed flutter margins for the simulation.

Time is not a commanded variable in this simulation so the length of the flight test is determined by the computational analysis time. The only known function of time is the flight path between test points. The aircraft is instructed to increase the dynamic pressure by 100 lb/ft^2 and stabilize at Mach 1.2 in 30 seconds.

The flight path as determined by the dynamic pressure for Mach 1.2 during the entire simulation is given in Figure 13.2. The horizontal portions of the flight path indicate time spent at a test point for which the flight conditions do not change. The length of time at each test point is determined by the computational cost of modal validation and robust flutter margin analysis. The unequal computational times at different test points demonstrated in Figure 13.2 is due to variations in the number of iterations required to generate and validate increased uncertainty levels and compute the corresponding μ levels.

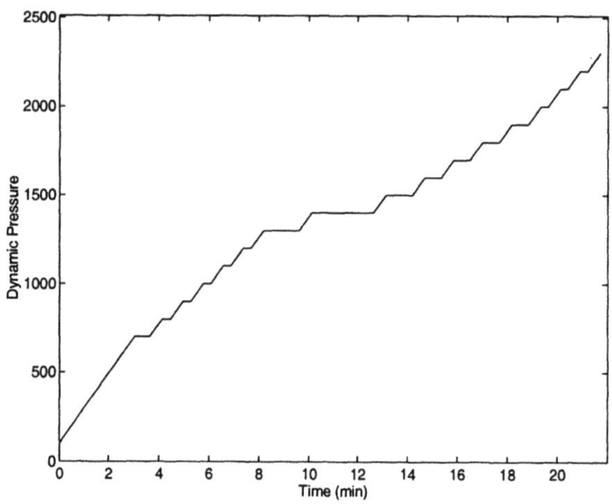

Fig. 13.2: Flight Path at Constant Mach=1.2

The true aircraft dynamics are time-varying due to decreasing mass of P_{true} throughout the simulation. The nominal and robust plant models used for flutter margin prediction are formulated for the heavyweight flight condition and thus are progressively worse representations of the true aircraft dynamics. Consequently, the amount of uncertainty associated with the robust model must increase to ensure the allowed range of dynamic variations includes the true dynamics. The initial mass is chosen as 95% of the full-fuel heavyweight condition and decreases at a rate of 5% of the heavyweight value for 20 minutes of flight time. The mass of P_{true} throughout the simulation is shown in Figure 13.3 and has no horizontal segments because mass is continuously decreasing whether the aircraft is at a test point or transitioning to a new test point.

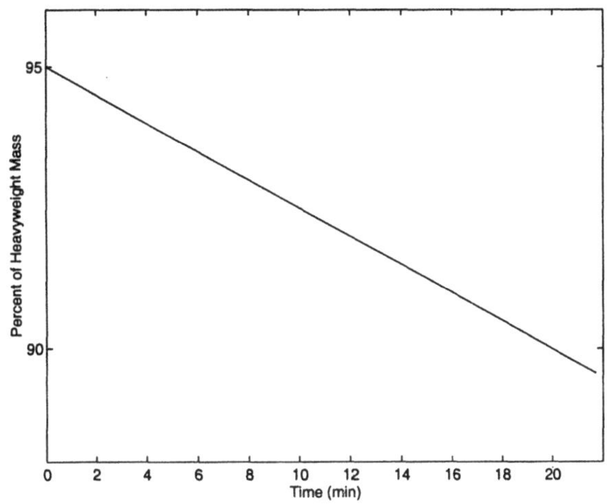

Fig. 13.3: Mass of the Aircraft P_{true} Expressed as Percentage of Heavyweight Condition

The time-varying mass will affect every aeroelastic modal response. The decrease in mass will appear as increases in natural frequency and damping. The modal characteristics for the trailing edge flap mode of P_{true} at $\overline{q} = 1000\ lb/ft^2$ are $\omega = 25.68\ Hz$ and $\zeta = .0125$ for 95% of heavyweight mass while $\omega = 25.74\ Hz$ and $\zeta = .0131$ for 90% of heavyweight mass. P_{rob} must be robust to these changes in modal dynamics and the corresponding change in flutter margins of P_{true}.

The uncertainty levels of P_{rob} are analyzed at each test point to ensure the measured flight data does not invalidate the model. If the robust model is invalidated, then the uncertainty weights are iteratively increased and analyzed until the amount of uncertainty is sufficient to account for the observed variations between P_{true} and P_{rob}. This simulation increases the amount of modal uncertainty to ensure the flight data could be generated by the robust model while keeping the amount of input multiplicative uncertainty constant.

This approach was used because it is anticipated the only time-varying errors in the model would be associated with specific parameters so the parametric uncertainty is the logical operator to update.

The actual time-varying levels of the parametric uncertainty are chosen automatically by the flutterometer tool and not influenced by any user interaction. These levels are chosen by iterating over perturbation magnitudes until the flight data at each test point does not invalidate the model. The flutterometer demonstrates the largest increases in uncertainty are associated with the dampings and smaller increases are associated with the natural frequencies. The amount of uncertainty in damping throughout the simulation is expressed as a percent of the nominal damping value in Figure 13.4.

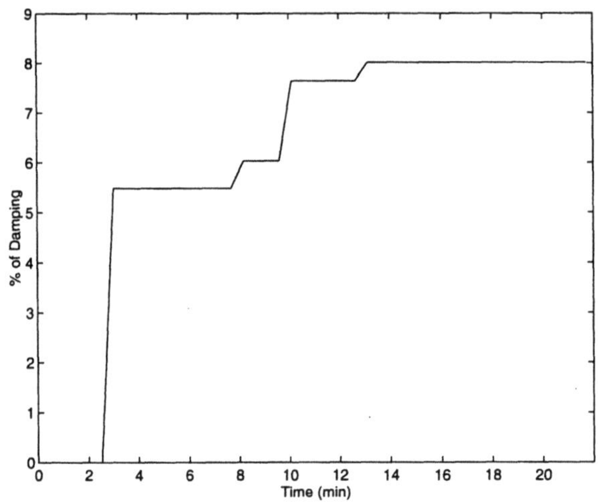

Fig. 13.4: Level of Modal Damping Uncertainty

The initial uncertainty levels in the simulation are pre-defined based on previous analysis of the F/A-18 SRA aircraft and models. Figure 13.4 demonstrates this initial uncertainty level of approximately 5% is sufficient to validate the robust model according to flight data recorded between 3 and 8 minutes of the simulation.

Increases in the damping uncertainty are generated at several points between 8 and 14 minutes of the simulation. The amount of these increases is determined solely by the model validation algorithm. The damping uncertainty levels do not increase after 19 minutes because the model validation procedure encounters some conditioning problems for the models that have damping values of less than 0.005 and are extremely close to an instability.

13.4 On-Line Robust Flutter Margins

Flutter margins are computed at each test point using the nominal model and the associated levels of uncertainty that are required to validate the model with flight data from that test point. Figure 13.5 presents these robust and nominal flutter margins along with the true flutter margin.

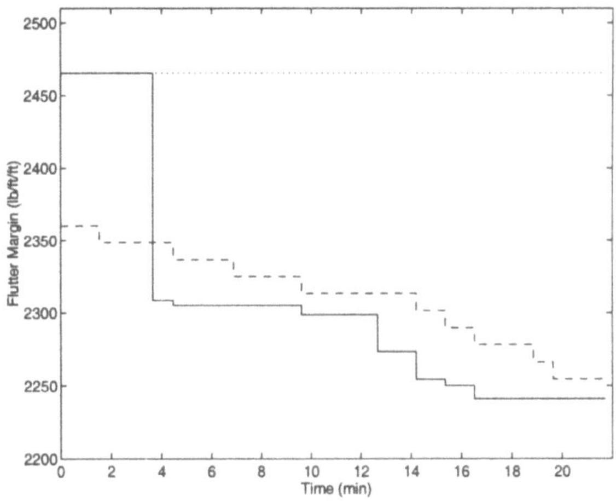

Fig. 13.5: Flutter Margins expressed as Dynamic Pressure Flight Condition : Robust μ Prediction (—), Nominal μ Prediction (...), True Flutter Margin (- - -)

The true flutter margin is time-varying and reflects the change in dynamics that results from the decreasing mass. The nominal flutter margin is initially close to the true flutter margin; however, the nominal plant does not account for time-varying mass so the accuracy of the nominal flutter margin steadily decreases. The robust flutter margin is initially constant and similar to the nominal flutter; however, the uncertainty description is altered after several test points so the robust flutter margin has a time-varying behavior.

The on-line implementation based on updates at test points allows the robust flutter margin to remain conservative with respect to the time-varying true flutter margin. This conservatism is a direct result of the worst-case nature of the μ computation with respect to the uncertainty description. The uncertainty is chosen to ensure the robust model encompasses the dynamics of the true model at each point so the resulting robust margin should be no greater than the true margin. Note the robust margin is only conservative to the true margin at the test point from which data is measured because the uncertainty description can not account for changes in dynamics resulting from future unknown time-varying effects that have not been observed.

The flutterometer is implemented in this simulation to demonstrate the information that would be available throughout the flight test from an on-line implementation of the μ method. This tool is implemented to describe the largest increase in dynamic pressure to which the aircraft model is robustly stable and to which the envelope may be safely expanded. Figure 13.6 displays the flutterometer readings throughout the simulation and clearly shows the on-line μ method continuously presents valuable information relating to the true flutter margin despite unmodeled time-varying dynamics.

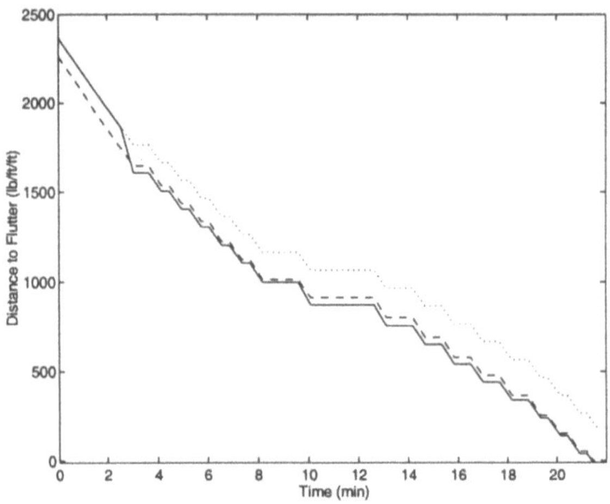

Fig. 13.6: Flutterometer showing Distance to Flutter in Dynamic Pressure : Robust μ Prediction (—), Nominal μ Prediction (...), True Distance to Flutter (- - -)

The benefits of the μ based flutterometer are easily seen when considering the traditional method of predicting flutter margins based on the evolution of modal damping as shown in Figure 13.7. Tracking modal damping is clearly inferior to the μ method for on-line flutter prediction. Modal damping provides little information until minute 17 when the aircraft is approaching flutter. After that time, damping shows a decreasing trend to indicate proximity to an instability; however, the nonlinear behavior precludes calculating the exact proximity of that instability.

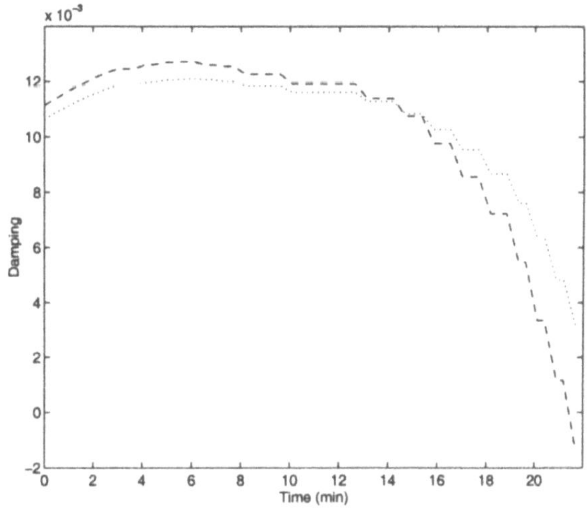

Fig. 13.7: Modal Damping for the Trailing Edge Flap Mode : Nominal Model P_{nom} (...), True Aircraft (- - -)

CHAPTER 14

CONCLUSION

Robust aeroservoelastic stability margins are analyzed by applying the μ method to an aircraft model with associated uncertainties. These margins indicate the smallest perturbation in dynamic pressure for which the model is not robustly stable to those uncertainties. Correspondingly, nominal stability margins consider only a theoretical plant and may be highly suspect because of errors and unmodeled dynamics. Thus, robust margins provide a higher level of safety and are confidence for determining a flight envelope within which an aircraft can operate.

The μ method presents a significant improvement over traditional methods of aeroservoelastic stability analysis because flight data can easily be incorporated into the analysis. This method uses the flight data to indicate the level of errors in a theoretical model and generate an updated model with associated uncertainty operators based on these errors. In this way, robust aeroservoelastic stability margins are computed that are worst-case with respect to the dynamics observed with flight data.

This method is particularly attractive for flight test programs and can be used to increase the efficiency of envelope expansion. Certainly the inherent ability to utilize flight data suggests an on-line implementation to predict margins during a flight. Furthermore, the μ parameter is used as the stability margin and its essentially linear variation across flight condition makes it a valid stability predictor to replace the stability indicator of damping which varies nonlinearly across flight condition.

REFERENCES

1. R.J. Adams, J.M. Buffington, A.G. Sparks and S.S. Banda, *Robust Multivariable Flight Control*, Advances in Industrial Control series, Springer-Verlag, London, 1994.
2. D. Afolabi, R.M.V. Pidaparti, and H.T.Y. Yang, "Flutter Prediction Using an Eigenvector Orientation Approach," *AIAA Journal*, Vol. 36, No. 1, January 1998, pp. 69-74.
3. P. Apkarian, J. Chretien, P. Gahinet and J. Biannic, "μ Synthesis by *D-K* Iterations with Constant Scalings," *American Control Conference*, San Francisco CA 1993, pp. 3192-3196.
4. K. Appa, "Constant Pressure Panel Method for Supersonic Unsteady Airloads Analysis," *Journal of Aircraft*, Vol. 24, No. 10, October 1987, pp. 696-702.
5. G.J. Balas, J.C. Doyle, K. Glover, A. Packard and R. Smith, *μ-Analysis and Synthesis Toolbox - Users Guide*, The MathWorks, Natick MA, 1991.
6. G.J. Balas and P.M. Young, "Control Design for Variations in Structural Natural Frequencies," *Journal of Guidance, Control, and Dynamics*, Vol. 18, No.2, March 1995, pp. 325-332.
7. G.A. Bécus, "Automated Search for the Most Critical Flutter Configuration in Models with Uncertainty," *Mathematical and Computer Modelling*, Vol. 14, 1990, pp. 977-982.
8. G.A. Bécus and C. Rekow, "Robust Control Theory Methods in Flutter Analysis," *International Forum on Aeroelasticity and Structural Dynamics*, Aachen, Germany, June 1991, pp. 464-468.
9. G.A. Bécus, "Stability Robustness of Aeroelastic Systems : A Perturbation Approach," *International Forum on Aeroelasticity and Structural Dynamics*, Rome, Italy, June 1997, Volume 2, pp. 425-532.
10. R.M. Bennett, *Application of Zimmerman Flutter-Margin Criterion to a Wind-Tunnel Model*, NASA-TM-84545, November 1982.
11. A. Berman and E.J. Nagy, "Improvement of a Large Analytical Model Using Test Data," *AIAA Journal*, Vol. 21, No. 8, 1993, pp. 1168-1173.
12. R.L. Bisplinghoff and H. Ashley, *Principles of Aeroelasticity*, Dover Publications, Inc., New York, 1962.
13. R.L. Bisplinghoff, H. Ashley, and R.L. Halfman, *Aeroelasticity*, Addison-Wesley Publishing, Cambridge MA, 1955.
14. M.A. Blair, *A Compilation of the Mathematics Leading to the Doublet Lattice Method*, Air Force Wright Laboratory, Dayton OH, WL-TR-92-3028, March 1992.
15. R.G. Bland, D. Goldfarb and M.J. Todd, "The Ellipsoid Method : A Survey," *Operations Research*, Vol. 28, No. 6, November-December 1981, pp. 1039-1091.
16. J.J. Block and T.W. Strganac, "Applied Active Control for a Nonlinear Aeroelastic Structure," *Journal of Guidance, Control, and Dynamics*, Vol. 21, No. 6, November-December 1998, pp. 838-845.
17. A. de Boer and M.H.M. Ellenbroek, "A Study on Methods to Compare Measured and Calculated Modal Data," *NLR Technical Paper*, National Aerospace Laboratory NLR, The Netherlands, NLR-TP-93095-U, March 1993.
18. S. Boyd and L. El Ghaoui, "Method of Centers for Minimizing Generalized Eigenvalues," *Linear Algebra and Applications : Special Issue on Numerical Linear Methods in Control*, July 1993, pp. 63-112.

19. S. Boyd, L. El Ghaoui, E. Feron and V. Balakrishnan, *Linear Matrix Inequalities in System and Control Theory*, Studies in Applied Mathematics, Society for Industrial and Applied Mathematics, Philadelphia PA, 1994.
20. R.P. Braatz, P.M. Young, J.C. Doyle and M. Morari, "Computational Complexity of μ Calculation," *IEEE Transactions on Automatic Control*, Vol. 39, No. 5, May 1994, pp. 1000-1002.
21. M.J. Brenner, "Actuator and Aerodynamic Modeling for High-Angle-of-Attack Aeroservoelasticity", April 1993, AIAA-93-1419.
22. M.J. Brenner, *Aeroservoelastic Modeling and Validation of a Thrust-Vectoring F/A-18 Aircraft*, NASA-TP-3647, September 1996.
23. M. Brenner and R. Lind, "Wavelet Filtering to Reduce Conservatism in Aeroservoelastic Robust Stability Margins," *AIAA Structures, Structural Dynamics, and Materials Conference*, Long Beach CA, April 1998, AIAA-98-1896.
24. M. Brenner and R. Lind, "Wavelet-Processed Flight Data for Robust Aeroservoelastic Stability Margins," *Journal of Guidance, Control and Dynamics*, Vol. 21, No. 6, November-December 1998, pp. 823-829
25. M. Brenner and R. Lind, "On-Line Aeroelastic Robust Stability Prediction using Wavelet Filtering," 21st *Congress of the International Council of Aeronautical Sciences*, Melbourne, Australia, September 13-18 1998, ICAS-98-4.9.1.
26. M. Brenner, R. Lind and D. Voracek, "Overview of Recent Flight Flutter Testing Research at NASA Dryden," *AIAA Structures, Structural Dynamics, and Materials Conference*, Orlando FL, April 1997, AIAA-97-1023.
27. C. S. Buttrill, B. B. Bacon, J. H. Heeg, J. A. Houck, and D. V. Wood, *Aeroservoelastic Simulation of an Active Flexible Wing Wind Tunnel Model*, NASA-TP-3510, April 1996.
28. J. Chen, "Frequency-Domain Tests for Validation of Linear Fractional Uncertain Models," *IEEE Transactions on Automatic Control*, Vol. 42, No. 6, June 1997, pp. 748-760.
29. T. Chen and L. Qiu, "$\mathcal{H}_\infty$ Design of Multirate Sampled-Data Control Systems", *Automatica*, Vol. 30, 1994, pp. 1139-1152.
30. J.E. Cooper and T.E. Noll, "Technical Evaluation Report on the 1995 Specialists Meeting on Advanced Aeroservoelastic Testing and Data Analysis," *Proceedings of the* 80th *AGARD Structures and Materials Panel*, AGARD-CP-566, Rotterdam, The Netherlands, May 8-10 1995, pp. T1-T10.
31. T. J. Cowan, A. S. Arena, Jr., and K. K. Gupta, "Accelerating CFD-Based Aeroelastic Predictions Using System Identification", AIAA-98-4152.
32. J.C. Doyle, B.A. Francis and A.R. Tannenbaum, *Feedback Control Theory*, MacMillan Publishing, New York, NY, 1992.
33. E.H. Dowell, "Nonlinear Aeroelasticity," *AIAA Structures, Structural Dynamics, and Materials Conference*, Long Beach CA, April 1990, AIAA-90-1031, pp. 1497-1509.
34. E. H. Dowell, E. F. Crawley, H. C. Curtiss, Jr., D. A. Peters,R. H. Scanlan, and F. Sisto, *A Modern Course in Aeroelasticity*, Kluwer Academic Publishers, The Netherlands, 1995.
35. E.H. Dowell and M. Ilgamov, *Studies in Nonlinear Aeroelasticity*, Springer-Verlag, New York, NY, 1988.
36. L.G. Duchesne, *Advanced Techniques for Flutter Clearance*, M.S. Thesis, Massachusetts Institute of Technology, Department of Aeronautics and Astronautics, January 1997.
37. J. W. Edwards, *Analysis of an Electrohydraulic Aircraft Control-surface Servo and Comparisons with Test Results*, NASA-TN-D-6928, August 1972.
38. D.F. Enns, *Model Reduction for Control System Design*, Ph.D. Thesis, Stanford University, 1984.

39. M.K.H. Fan, A.L. Tits and J.C. Doyle, "Robustness in the Presence of Mixed Parametric Uncertainty and Unmodeled Dynamics," *IEEE Transactions on Automatic Control*, Vol. 36, No. 1, January 1991, pp. 25-38.
40. L.R. Felt, L.J. Huttsell, T.E. Noll, and D.E. Cooley, "Aeroservoelastic Encounters", *Journal of Aircraft*, Vol. 16, No. 7, July 1979, pp. 477-483.
41. E. Feron, "A More Reliable Robust Stability Indicator for Linear Systems Subject to Parametric Uncertainties," *IEEE Transactions on Automatic Control*, Vol. 42, No. 9, September 1997, pp. 1326-1330.
42. E. Feron, "Robustness of Linear Systems Against Parametric Uncertainties: Towards Consistent Stability Indicators," *IEEE Conference on Decision and Control*, New Orleans LA, December 1995, pp. 1425-1430.
43. A. Feuer and G. C. Goodwin, *Sampling in Digital Signal Processing and Control*, Birkhauser Boston, Cambridge, Mass., 1996.
44. L.C. Freudinger, R. Lind and M. Brenner, "Correlation Filtering of Modal Dynamics using the Laplace Wavelet," *International Modal Analysis Conference*, Santa Barbara CA, February 1998, pp. 868-877.
45. P. Gahinet, A. Nemirovski, A.J. Laub and M. Chilali, *LMI Control Toolbox*, The MathWorks, Natick MA, 1995.
46. D.R. Gaukroger, C.W. Skingle and K.H. Heron, "An Application of System Identification to Flutter Testing," *Journal of Sound and Vibration*, Vol. 72, No. 2, 1980, pp. 141-150.
47. A. Gelb and W.E. Vander Velde, *Multiple-Input Describing Functions and Nonlinear System Design*, McGraw-Hill Book Company, New York, 1968.
48. J.P. Giesing, T. Kalman and W. Rodden, *Subsonic Unsteady Aerodynamic for General Lattice Method, Part I - Vol I - Direct Application of the Nonplanar Doublet Lattice Method*, AFFDL-TF-71-5, November 1971.
49. W. Gilbert, L. Nguyen and J. Gera, "Control Law Research in the NASA High Alpha Technology Program," AGARD CP-465, April 1991.
50. H.C. Gilliatt, T.W. Strganac, and A.J. Kurdila, "Nonlinear Aeroelastic Response of an Airfoil," *AIAA Aerospace Sciences Meeting*, Reno NV, January 1997, AIAA-97-0459.
51. K.D. Gondoly, *Application of Advanced Robustness Analysis to Experimental Flutter*, M.S. Thesis, Massachusetts Institute of Technology, Department of Aeronautics and Astronautics, June 1995.
52. M. Green and D.J.N. Limebeer, *Linear Robust Control*, Prentice-Hall Inc., Englewood Cliffs, NJ, 1995.
53. K.K. Gupta, STARS - *A General-Purpose Finite Element Computer Program for Analysis of Engineering Structures*, NASA-TP-1129, 1984.
54. K.K. Gupta, STARS - *An Integrated General Purpose Finite Element Structural, Aeroelastic, and Aeroservoelastic Analysis Computer Program*, NASA-TM-101709, 1990.
55. K. K. Gupta, "Development of a Finite Element Aeroelastic Analysis Capability", *Journal of Aircraft*, Vol. 33, No. 5, September-October 1996, pp. 995-1002.
56. K.K. Gupta, M.J. Brenner and L.S. Voelker, *Development of an Integrated Aeroservoelastic Analysis Program and Correlation with Test Data*, NASA-TP-3120, May 1991.
57. H.J. Hassig, "An Approximate True Damping Solution of the Flutter Equation by Determinant Iteration," *Journal of Aircraft*, Vol. 8, No. 11, November 1971, pp. 885-889.
58. J. He and D.J. Ewins, "Compatibility of Measured and Predicted Vibration Modes in Model Improvement Studies," *AIAA Journal*, Vol. 29, No. 5, September-October 1991, pp. 798-803.

59. J.P. How and S.R. Hall, "Connections Between the Popov Stability Criterion and Bounds for Real Parameter Uncertainty," *American Control Conference*, San Francisco CA, June 1993, pp. 1084-1089.
60. P.M. Hutin, "State of the Art and Open Problems in Aeroelasticity," presented at *International Forum on Aeroelasticity and Structural Dynamics*, Rome, Italy, June 1997.
61. T. Iwasaki, "Robust Performance Analysis for Systems with Norm-Bounded Time-Varying Structured Uncertainty," *American Control Conference*, Baltimore MD, June 1994, pp. 2343-2347.
62. K.E. Kadrnka, "Multimode Instability Prediction Method," *AIAA Structures, Structural Dynamics, and Materials Conference*, Orlando FL, April 1985, AIAA-85-0737, Volume 2, pp. 435-442.
63. M. Karpel, "Design for the Active Flutter Suppression and Gust Alleviation Using State-Space Aeroelastic Modeling," *Journal of Aircraft*, Vol. 19, March 1982, pp. 221-227.
64. M. Karpel and E. Strul, "Minimum-State Unsteady Aerodynamic Approximations with Flexible Constraints", *Journal of Aircraft*, Vol. 33, No. 6, November-December 1996, pp. 1190-1196.
65. H. Katz, F.G. Foppe, and D.T. Grossman, "F-15 Flight Flutter Test Program," *Flutter Testing Techniques*, NASA-SP-415, October 1975, pp. 413-431.
66. M.W. Kehoe, "A Historical Overview of Flight Flutter Testing," *Proceedings of the* 80th *AGARD Structures and Materials Panel*, AGARD-CP-566, Rotterdam, The Netherlands, May 8-10 1995, pp. 1-15.
67. J. Ko, A.J. Kurdila and T.W. Strganac, "Nonlinear Control of a Prototypical Wing Section with Torsional Nonlinearity," *Journal of Guidance, Control, and Dynamics*, Vol. 20, No. 6, November-December 1997, pp. 1181-1189.
68. J. Ko, A.J. Kurdila and T.W. Strganac, "Nonlinear Control Theory for a Class of Structural Nonlinearities in a Prototypical Wing Section," *AIAA Aerospace Sciences Meeting*, Reno NV, January 1997, AIAA-97-0580.
69. J. Ko, T.W. Strganac, and A.J. Kurdila, "Stability and Control of a Structurally Nonlinear Aeroelastic System," *Journal of Guidance, Control, and Dynamics*, Vol. 21, No. 4, September-October 1997, pp. 718-725.
70. A. Kumar and G.J. Balas, "An Approach to Model Validation in the μ Framework," *American Control Conference*, Baltimore MD, June 1994, pp. 3021-3026.
71. R.M. Lin and M.K. Lim, "Analytical Model Updating and Model Reduction," *AIAA Journal*, Vol. 34, No. 9, 1996, pp. 1966-1969.
72. R. Lind, *A Presentation on Robust Flutter Margin Analysis and a Flutterometer*, NASA-TM-97-206220, October 1997.
73. R. Lind, G.J. Balas and A. Packard, "Robustness Analysis with Linear Time-Invariant and Time-Varying Real Uncertainty," *AIAA Guidance, Navigation and Control Conference*, July 1995, AIAA-95-3188, pp 132-140.
74. R. Lind and M. Brenner, "Worst-Case Flutter Margins from F/A-18 Aircraft Aeroelastic Data," *AIAA Structures, Structural Dynamics, and Materials Conference*, Orlando FL, April 1997, AIAA-97-1266.
75. R. Lind and M. Brenner, "Robust Flutter Margins of an F/A-18 Aircraft from Aeroelastic Flight Data," *Journal of Guidance, Control and Dynamics*, Vol. 20, No. 3, May-June 1997, pp. 597-604.
76. R. Lind and M. Brenner, "A Worst-Case Approach for On-Line Flutter Prediction," *International Forum on Aeroelasticity and Structural Dynamics*, Rome, Italy, June 1997, Volume 2, pp. 79-86.
77. R. Lind and M. Brenner, *Robust Flutter Margin Analysis that Incorporates Flight Data*, NASA-TP-1998-206543, March 1998.

78. R. Lind and M. Brenner, "Analyzing Aeroservoelastic Stability Margins using the μ Method," *AIAA Structures, Structural Dynamics and Materials Conference*, Long Beach CA, April 1998, AIAA-98-189.
79. R. Lind and M. Brenner, "Incorporating Flight Data into a Robust Aeroelastic Model," *Journal of Aircraft*, Vol.35, No. 3, May-June 1998, pp. 470-477.
80. R. Lind, M. Brenner and L.C. Freudinger, "Improved Flight Test Procedures for Flutter Clearance," *International Forum on Aeroelasticity and Structural Dynamics*, Rome, Italy, June 1997, Volume 3, pp. 291-298.
81. R. Lind, L.C. Freudinger and D.F. Voracek, "A Comparison of Aeroelastic Excitation Mechanisms," *Journal of Aircraft*, Vol. 35, No. 5, September-October 1998, pp. 830-832.
82. R. Lind, K. Snyder and M. Brenner, "Investigating Transient and Limit Cycle Behaviors of a Nonlinear Structure by Wavelet Transforms," *AIAA Structures, Structural Dynamics and Materials Conference*, Long Beach CA, April 1998, AIAA-98-1808.
83. E. Livne, "Integrated Aeroservoelastic Optimization : Status and Direction," *AIAA Structures, Structural Dynamics and Materials Conference*, Orlando FL, April 1997, AIAA-97-1409.
84. L. Ljung, *System Identification : Theory for the User*, Prentice-Hall Inc., Englewood Cliffs, NJ, 1987.
85. L. Ljung and L. Guo, "The Role of Model Validation for Assessing the Size of Unmodeled Dynamics," *IEEE Transactions on Automatic Control*, Vol. 42, No. 9, September 1997, pp. 1230-1239.
86. J.M. Maciejowski, *Multivariable Feedback Design*, Addison Wesley Publishers Ltd, England, 1989.
87. Y. Matsuzaki and Y. Ando, "Estimation of Flutter Boundary from Random Responses Due to Turbulence at Subcritical Speeds," *Journal of Aircraft*, Vol. 18, No. 10, October 1981, pp. 862-868.
88. Y. Matsuzaki and Y. Ando, "Divergence Boundary Prediction from Random Responses: NAL's Method," *Journal of Aircraft*, Vol. 21, No. 6, June 1984, pp. 435-436.
89. Y. Matszaki and Y. Ando, "Flutter and Divergence Boundary Prediction from Nonstationary Random Responses at Increasing Speds," *AIAA Structures, Structural Dynamics, and Materials Conference*, Orlando FL, April 1985, AIAA-85-0691, Volume 2, pp. 313-320.
90. D. McRuer, I. Ashkenas, and D. Graham, *Aircraft Dynamics and Automatic Control*, Princeton University Press, 1973.
91. L. Morino, F. Mastroddi, R. De Troia, G.L. Ghiringhelli and P. Mantegazza, "Matrix Fraction Approach for Finite-State Aerodynamic Modeling," *AIAA Journal*, Vol. 33, No. 4, April 1995, pp. 703-711.
92. L. Morino, F. Mastroddi, R. De Troia and M. Pecora, "On Finite-State Aerodynamic Modeling with Applications to Aeroservoelasticity," *Proceedings of Forum International Aeroelasticite et Dynamique de Structures*, Association Aeronautique et Astronautique de France, Volume 1, 1993, pp. 97-116.
93. V. Nalbantoğlu, G.J. Balas and P. Thomson, "The Role of Performance Criteria Selection in the Control of Flexible Structures," *AIAA Guidance, Navigation and Control Conference*, San Diego CA, July 1996, AIAA-96-3841.
94. E. Nissim, "Reduction of Aerodynamic Augmented States in Active Flutter Suppression Systems," *Journal of Aircraft*, Vol. 28, No. 1, January-February 1991, pp. 82-93.
95. E. Nissim and G.B. Gilyard, *Method of Experimental Determination of Flutter Speed by Parameter Identification*, NASA-TP-2923, 1989.

96. T. O'Neil and T.W. Strganac, "Nonlinear Aeroelastic Response - Analysis and Experiments," *AIAA Aerospace Sciences Meeting*, Reno NV, January 1996, AIAA-96-0014.
97. T. O'Neil, H.C. Gilliat and T.W. Strganac, "Investigations of Aeroelastic Response for a System with Continuous Structural Nonlinearities," *AIAA Structures, Structural Dynamics, and Materials Conference*, Salt Lake City UH, April 1996, AIAA-96-1390.
98. A. Packard and J. Doyle, "The Complex Structured Singular Value," *Automatica*, Vol. 29, No. 1, 1993, pp. 71-109.
99. J. Pahle, B. Powers, V. Regenie, V. Chacon, S. Degroote and S. Murnyack, *Research Flight Control System Development for the F-18 High Alpha Research Vehicle*, NASA-TM-104232, 1991.
100. E. Pendleton, D. Bessette, P. Field, G. Miller, and K. Griffin, "The Active Aeroelastic Wing Flight Research Program", AIAA-98-1972.
101. L. D. Peterson and E. F. Crawley, "Improved Exponential Time Series Approximation od Unsteady Aerodynamic Operators", *Journal of Aircraft*, Vol. 25, No. 2, February 1988, pp. 121-127.
102. D. M. Pitt and C. E. Goodman, "Flutter Calculations Using Doublet Lattice Aerodynamics Modified by the Full Potential Equations", AIAA-87-0882.
103. F. Poirion, "Impact of Random Uncertainties on Aircraft Aeroelastic Stability," *Conference on Stochastic Structural Dynamics*, San Juan, Porto Rico, January 1995.
104. F. Poirion, "Multi Mach State Space Modeling," *International Forum on Aeroelasticity and Structural Dynamics*, Rome, Italy, June 1997, Volume 2, pp. 307-311.
105. K. Poolla, P. Khargonekar, A. Tikku, J. Krause and K. Nagpal, "A Time-Domain Approach to Model Validation," *IEEE Transactions on Automatic Control*, Vol. 39, May 1994, pp. 951-959.
106. S.J. Price and B.H.K. Lee, "Development and Analysis of Flight Flutter Prediction Methods," *AIAA Dynamics Specialists Conference*, Dallax TX, April 1992, AIAA-92-2101, pp. 188-200.
107. S.J. Price and B.H.K. Lee, "Evaluation and Extension of the Flutter-Margin Method for Flight Flutter Prediction," *Journal of Aircraft*, Vol. 30, No. 3, May-June 1993, pp. 395-402.
108. A. Rantzer, "Uncertainties with Bounded Rates of Variations," *American Control Conference*, San Francisco CA, June 1993, pp. 29-30.
109. L.R. Ray and R.F. Stengel, "Application of Stochastic Robustness to Aircraft Control Systems," *Journal of Guidance, Controls and Dynamics*, Vol. 14, No. 6, November-December 1991, pp. 1251-1259.
110. W. Reed III, "A New Flight Flutter Excitation System," *Proceedings of the* 19th *Symposium*, Society of Flight Engineers, Arlington TX, August 14-18 1988.
111. R. Roberts, *Rational Functional Approximations to Unsteady Aerodynamics*, Ph.D. Thesis, University of Bristol, Department of Aerospace Engineering, October 1996.
112. E. A. Rodrigues and T. A. Weisshar, "Linear Discrete-Time State-Space Modeling of Dynamic Systems for Control Applications", AIAA-95-3304.
113. E. A. Rodrigues and T. A. Weisshar, "Nonlinear Discrete-Time State-Space Modeling of Dynamic Systems for Control Applications", AIAA-95-3305.
114. K.L. Roger, "Airplane Math Modeling Methods for Active Control Design," *Proceedings of the* 44th *AGARD Structures and Materials Panel*, AGARD-CP-228, Lisbon, Portugal, April 1977, pp. 4.1-4.11.
115. R. Roy and R. Walker, *Real-Time Flutter Identification*, NASA-CR-3933, October 1985.

116. C.L. Ruhlin, J.J. Watson, R.H. Ricketts, and R.V. Doggett, "Evaluation of Four Subcritical Response Methods for On-Line Prediction of Flutter Onset in Wind Tunnel Tests," *Journal of Aircraft*, Vol. 20, No. 10, October 1983, pp. 835-840.
117. J.S. Shamma, "Robust Stability with Time-Varying Structured Uncertainty," *IEEE Transactions on Automatic Control*, Vol. 39, No. 4, April 1994, pp. 714-724.
118. S.J. Shelley, L.C. Freudinger and R.J. Allemang, "Development of an On-Line Parameter Estimation System Using the Discrete Modal Filter," *International Modal Analysis Conference*, San Diego CA, February 1992, pp. 173-183.
119. S.J. Shelley and C.R. Pickrel, "New Concepts for Flight Flutter Parameter Estimation," *International Modal Analysis Conference*, Orlando FL, February 1997, pp. 490-496.
120. J.R. Sitz, "The F-18 Systems Research Aircraft Facility", *Aerotech Conference*, Anaheim CA, SAE Technical Paper 922063, 1992.
121. J.R. Sitz, *The F-18 Systems Research Aircraft Facility*, NASA-TM-4433, December 1992.
122. R.S. Smith and J.C. Doyle, "Model Validation : A Connection Between Robust Control and Identification," *IEEE Transactions on Automatic Control*, Vol. 37, No. 7, 1992, pp. 942-952.
123. K. Spaans, "Implementation and Verification of Parametric Finite Element Model Updating," *Fokker Space and Systems Technical Paper*, FSS-R-92-0270, August 1992.
124. R. Stirling, "Linearized Model of the Jaguar Fly-by-Wire Actuation System", University of Bristol, Department of Aeronautical Engineering, Report No. RS/2/83, October 1983.
125. R. Stirling, "Flexible Aircraft and Control System Dynamics", University of Bristol, Department of Aeronautical Engineering, Report No. RS/3/84, July 1984.
126. R. Stirling and D. A. Cowling, "Implementation of Comprehensive Actuation System Models in Aeroservoelastic Analysis", *European Forum on Aeroelasticity and Structural Dynamics,* Aachen, Germany. Report No. SDL 143, April 1989.
127. T.W. Strganac, H. Gilliatt and A.J. Kurdila, "On the Presence of Internal Resonances in Aeroelastic Systems," *AIAA Structures, Structural Dynamics, and Materials Conference*, Long Beach CA, April 1998, AIAA-98-1954.
128. S. H. Tiffany and W. M. Adams, Jr., *Nonlinear Programming Extensions to Rational Function Approximation Methods for Unsteady Aerodynamic Forces,* NASA-TP-2776, July 1988.
129. S. H. Tiffany and M. Karpel, "Aeroservoelastic Modeling and Applications Using Minimum-State Approximations of the Unsteady Aerodynamics", AIAA-89-1188.
130. J.M.T. Thomson and H.B. Steward, *Nonlinear Dynamics and Chaos*, John Wiley and Sons, Chichester England, 1986.
131. H. Torii and Y. Matsuzaki, "Flutter Boundary Prediction Based on Nonstationary Data Measurement," *Journal of Aircraft*, Vol. 34, No. 3, May-June 1997, pp. 427-432.
132. L. Vernon, *In-Flight Investigation of a Rotating Cylinder-Based Structural Excitation System for Flutter Testing*, NASA-TM-4512, June 1993.
133. R. Walker and N. Gupta, *Real-Time Flutter Analysis*, NASA-CR-170412, March 1984.
134. R. F. Whitbeck and L. G. Hofmann, "Analysis of Digital Flight Control Systems with Flying Qualities Applications", AFFDL-TR-78-115, September 1978.

135. R. F. Whitbeck, D. G. L. Didaleusky, and L. G. Hofmann, "Frequency Response of Digitally Controlled Systems", *AIAA Journal of Guidance, Control, and Dynamics*, Vol. 4, No. 4, July-August 1981, pp. 423-427.
136. C. D. Wieseman, "Methodology for Matching Experimental and Analytical Aerodynamic Data", April 1988, AIAA-88-2392.
137. S. Wiggins, *Introduction to Applied Nonlinear Dynamical Systems and Chaos*, Springer-Verlag, New York NY, 1990.
138. B. A. Winther, P. J. Goggin, and J. R. Dykman,, "Reduced Order Dynamic Aeroelastic Model Development and Integration with Nonlinear Simulation", AIAA-98-1897.
139. R. W. Wlezien, G. C. Horner, A. R. McGowan, S. L. Padula, M. A. Scott, R. J. Silcox, and J. O. Simpson, "The Aircraft Morphing Program", AIAA-98-1927.
140. Y. Yamamoto and P. R. Khargonekar, "Frequency Response of Sampled-Data Systems", *IEEE Transactions on Automatic Control*, Vol. 41, No. 2, February 1996, pp. 166-176.
141. P.M. Young, M.P. Newlin and J.C. Doyle, "μ Analysis with Real Parametric Uncertainty," *IEEE Conference on Decision and Control*, Brighton, England, December 1991, pp. 1251-1256.
142. G. Zames, "On the Input-Output Stability of Nonlinear Time-Varying Feedback Systems, Parts I and II," *IEEE Transactions on Automatic Control*, Vol. 11, 1966, pp. 228-238 and pp. 465-476.
143. Y. Zeng and S.N. Singh, "Output Feedback Variable Structure Adaptive Control of an Aeroelastic System," *Journal of Guidance, Control, and Dynamics*, Vol. 21, No. 6, November-December 1998, pp. 830-837.
144. Q. Zhang, R.J. Allemang and D.L. Brown, "Modal Filter : Concept and Applications," *International Modal Analysis Conference*, Orlando FL, January 1990, pp. 487-496.
145. H. Zimmerman, "Aeroservoelasticity", *Computer Methods in Applied Mechanics and Engineering*, Vol. 90, 1991, pp. 719-735.
146. N.H. Zimmerman and J.T. Weissenburger, "Prediction of Flutter Onset Speed Based on Flight Testing at Subcritical Speeds," *Journal of Aircraft*, Vol. 1, No. 4, July-August 1964, pp.190-202.

INDEX

MIX
Papier aus verantwortungsvollen Quellen
Paper from responsible sources
FSC® C105338

If you have any concerns about our products,
you can contact us on
ProductSafety@springernature.com

In case Publisher is established outside the EU,
the EU authorized representative is:
Springer Nature Customer Service Center GmbH
Europaplatz 3, 69115 Heidelberg, Germany

Printed by Libri Plureos GmbH
in Hamburg, Germany